Assessing the Conservation Value of Fresh Waters

Fresh water has many uses, such as for drinking, hydroelectric power and recreation. This creates conflict between conservation and exploitation. This book explores various aspects of conservation evaluation, including the selection of important areas for protection, responding to threats from catchment development and determining the restoration potential of degraded water bodies. Aimed at academic researchers, graduate students and professionals, chapters are written by pairs of UK and US authors, who compare methods used for evaluating rivers and lakes for conservation in these countries that share a long history of freshwater science, but approach nature conservation very differently. Sweden, Australia and South Africa are also examined, and there is a chapter on developing countries, allowing discussion of the role of social and economic conditions in conservation ethics.

PHILIP J. BOON works in the Policy and Advice Directorate at Scottish Natural Heritage. He is founder and Chief Editor of the journal *Aquatic Conservation: Marine and Freshwater Ecosystems*.

CATHERINE M. PRINGLE is Research Professor in the Odum School of Ecology at the University of Georgia (UGA). She serves as Chair of the Conservation Ecology Graduate Program at UGA and has been a recipient of their Creative Research Medal.

ECOLOGY, BIODIVERSITY AND CONSERVATION

The world's biological diversity faces unprecedented threats. The urgent challenge facing the concerned biologist is to understand ecological processes well enough to maintain their functioning in the face of the pressures resulting from human population growth. Those concerned with the conservation of biodiversity and with restoration also need to be acquainted with the political, social, historical, economic and legal frameworks within which ecological and conservation practice must be developed. The new Ecology, Biodiversity and Conservation series will present balanced, comprehensive, up-to-date, and critical reviews of selected topics within the sciences of ecology and conservation biology, both botanical and zoological, and both 'pure' and 'applied'. It is aimed at advanced final-year undergraduates, graduate students, researchers and university teachers, as well as ecologists and conservationists in industry, government and the voluntary sectors. The series encompasses a wide range of approaches and scales (spatial, temporal, and taxonomic), including quantitative, theoretical, population, community, ecosystem, landscape, historical, experimental, behavioural and evolutionary studies. The emphasis is on science related to the real world of plants and animals rather than on purely theoretical abstractions and mathematical models. Books in this series will, wherever possible, consider issues from a broad perspective. Some books will challenge existing paradigms and present new ecological concepts, empirical or theoretical models, and testable hypotheses. Other books will explore new approaches and present syntheses on topics of ecological importance.

The Ecology of Phytoplankton
C.S. Reynolds

Invertebrate Conservation and Agricultural Ecosystems
T.R. New

Risks and Decisions for Conservation and Environmental Management
Mark Burgman

Nonequilibrium Ecology
Klaus Rohde

Ecology of Populations
Esa Ranta, Veijo Kaitala and Per Lundberg

Ecology and Control of Introduced Plants
Judith H. Myers and Dawn Bazely

Systematic Conservation Planning
Chris Margules and Sahotra Sarkar

Assessing the Conservation Value of Fresh Waters
Philip J. Boon and Catherine M. Pringle

Bird Conservation and Agriculture
Jeremy D. Wilson, Andrew D. Evans and Philip V. Grice

Large-Scale Landscape Experiments
David B. Lindemayer

Assessing the Conservation Value of Fresh Waters:

An International Perspective

Edited by

PHILIP J. BOON

Scottish Natural Heritage, Edinburgh, UK

and

CATHERINE M. PRINGLE

Odum School of Ecology, University of Georgia, Athens, GA, USA

CAMBRIDGE UNIVERSITY PRESS
Cambridge, New York, Melbourne, Madrid, Cape Town, Singapore, São Paulo, Delhi

Cambridge University Press
The Edinburgh Building, Cambridge CB2 8RU, UK

Published in the United States of America by Cambridge University Press, New York

www.cambridge.org
Information on this title: www.cambridge.org/9780521848855

First published 2009

Printed in the United Kingdom at the University Press, Cambridge

A catalogue record for this publication is available from the British Library

Library of Congress Cataloguing in Publication data
Assessing the conservation value of freshwaters / edited by Philip J Boon, Catherine M Pringle.
p. cm.
1. Freshwater biodiversity conservation. 2. Freshwater ecology. I. Boon, P. J.
II. Pringle, Catherine M. III. Title.
QH96.8.B53A88 2008
333.95′2816–dc22

2008025659

ISBN 978-0-521-84885-5 hardback
ISBN 978-0-521-61322-4 paperback

Contents

Contributors

ROBIN ABELL
Conservation Science Program, World Wildlife Fund-US, Washington, USA

COLIN W. BEAN
Scottish Natural Heritage, Clydebank, UK

PHILIP J. BOON
Scottish Natural Heritage, Edinburgh, UK

ANDREW BOULTON
Ecosystem Management, University of New England, Armidale, Australia

MARK BRYER
The Nature Conservancy, Bethesda, MD, USA

CATHERINE DUIGAN
Countryside Council for Wales, Bangor, UK

LAURIE DUKER
LakeNet, Annapolis, USA

MARY FREEMAN
US Geological Survey, University of Georgia, Athens, USA

C.A. FRISSELL
Pacific Rivers Council, Polson, USA

JONATHAN HIGGINS
The Nature Conservancy, Chicago, USA

T.E.L. LANGFORD
Centre for Environmental Sciences, University of Southampton, Southampton, UK.

JON NEVILL
OnlyOnePlanet Consulting, Hampton, Australia

JAY O'KEEFFE
Institute for Water Research, Rhodes University, Grahamstown, South Africa

MARGARET PALMER
Nethercott, Stamford Road, Barnack, UK

CATHERINE M. PRINGLE
Odum School of Ecology, University of Georgia, Athens, USA

CHRISTA THIRION
Resource Quality Services, Department of Water Affairs and Forestry, Pretoria, South Africa

EVA WILLÉN
Swedish University of Agricultural Sciences, Department of Environmental Assessment, Uppsala, Sweden

DAVID WITHRINGTON
Natural England, Peterborough, UK

1 · *Introduction*

PHILIP J. BOON AND CATHERINE M. PRINGLE

Challenges facing freshwater ecosystems are immense. Moreover, we have entered an unprecedented era of globalization, climate change and increased urban development where these factors are interacting (often in ways that we do not yet completely understand) to undermine the integrity of freshwater ecosystems. The global human population may reach 10 billion by 2050, with increasing demands for freshwater resources and consequent negative effects on lakes and streams.

On an encouraging note, freshwater resource and conservation issues are finding their way more and more into the public consciousness, with a rapid increase in available information on freshwater conservation relative to general nature conservation (Table 1.1). International political recognition of the global water crisis is evidenced by the formation of the World Water Council in 1996. Its mission is 'to promote awareness, build political commitment and trigger action on critical water issues at all levels and to facilitate the efficient conservation, protection, development, planning, management and use of water in all its dimensions on an environmentally sustainable basis for the benefit of all life on earth' (www.worldwatercouncil.org). The World Water Forum (which is a product of the World Water Council) is held every 3 years and provides a much-needed platform for international communication, with the goal of reaching a common strategic vision on water resources and water services management.

Despite this increasing public and international awareness of freshwater resource and conservation issues, conservation studies of freshwater ecosystems are often neglected or overlooked within the field of conservation science. For example, in a recent paper, Lawler *et al.* (2006) attempted to assess gaps in conservation research over a 20-year period. In this 20-year 'report card', based on 628 papers from 14 journals, it was concluded that only 21% of all conservation research covers aquatic ecosystems. This conclusion is misleading as the survey did not include many of the aquatic journals in which conservation-related papers are published, such as *Aquatic*

Assessing the Conservation Value of Fresh Waters, ed. Philip J. Boon and Catherine M. Pringle. Published by Cambridge University Press.

Table 1.1 *Number of websites containing conservation-related terms, found using the search engine 'Google.co.uk' in December 2007 compared with May 2005*

Search term	2005	2007	change (%)
Nature conservation	950 000	1 490 000	+57
Natural resource conservation	414 000	348 000	−16
Habitat conservation	373 000	527 000	+41
Species conservation	179 000	244 000	+36
River conservation	89 900	125 000	+39
Wetland conservation	89 500	116 000	+30
Aquatic conservation	35 700	48 200	+35
Lake conservation	23 300	48 200	+107
Freshwater conservation	955	14 000	+1366

Conservation, *Freshwater Biology*, *River Research and Applications*, *Journal of the North American Benthological Society*, *Limnology and Oceanography*, *Canadian Journal of Fisheries and Aquatic Sciences*, *Transactions of the American Fisheries Society*, *Fisheries* and *Journal of the American Water Resources Association*. Similarly, rivers have been largely omitted from the mainstream scientific literature that deals with the management and conservation of fragmented landscapes (but see Pringle, 2006; Crooks & Sanjayan, 2006). Thus, it is critical to communicate more effectively the importance of freshwater ecosystems and methods for assessing the value of fresh waters for conservation.

Towards that end, this book is about the process of assessing the value of fresh waters for nature conservation, and is an outgrowth of two other books on freshwater conservation by one of the editors (Boon *et al.*, 1992, 2000). There are many definitions of the term 'conservation', but in the context of this volume we understand it to mean the maintenance of natural resources through protection, careful management and, where necessary, rehabilitation or restoration. There will always be a degree of subjectivity in what is considered 'important' or 'valuable' in the area of freshwater conservation. However, unless there are some generally accepted protocols for assessing conservation value, it is difficult to make progress in selecting rivers or lakes for special protection, or in determining the relative merits of catchment development in one place as opposed to another. Incentives for freshwater conservation range from the practical to the philosophical. Freshwater ecosystems provide critical goods and services to human societies, including water purification, flood control and nutrient cycling (National Research Council, 2005). Nature can also be assigned value for its own sake – a recognition both of the 'rights' of habitats and species to exist and of their role in enriching human experience. This may be predominantly a 'first-world' view (see Chapter 12; Boon *et al.*, 2000; Wishart *et al.*, 2000), but extremely relevant where it is invoked.

This book takes an innovative approach to the assessment of fresh waters for their nature conservation value by examining the subject from both sides of the Atlantic. The UK and the USA, although very different in size, landscape, climatic variation, history and politics, have each had a long association with freshwater science. In contrast, nature conservation policy and practice are quite different in both countries. In the UK (i.e. England, Scotland, Wales, Northern Ireland), the conservation movement has developed through mainstream government agencies over more than half a century, influenced strongly by European legislation. The USA, while strongly influenced by national legislation, is clearly less influenced by international legislation, and non-governmental organizations (NGOs) have played a key role in the development of freshwater conservation for natural values (Chapter 3).

Most of the chapters in this book have been written by pairs of authors – each pair comprising one author from the UK and one from the USA. Compiling this volume has been an interesting and challenging experience, not least because authors and editors alike have discovered the limitations in their knowledge and understanding of how things work in each other's countries. The contributors to this book are drawn from a wide range of organizations – government conservation agencies, NGOs, universities, research institutes, consultancies – but all share in common a direct involvement in freshwater science and the conservation of freshwater habitats and species. In writing each chapter, the authors have followed some general principles set out by the editors. First and foremost, the book is about the assessment of 'natural values' of freshwater ecosystems – specifically habitats, biota and ecological processes – rather than wider aspects such as economic or aesthetic values. It covers the evaluation of rivers and lakes, rather than attempting to extend its scope to other freshwater systems such as ponds or wetlands. This is not due to a lack of recognition of their importance, but rather as a practical means of limiting the length of the book. To some extent it also reflects the present balance of freshwater work in the UK and the USA, especially in response to recent legislative imperatives such as the EC Water Framework Directive (in the case of the UK).

A brief 'road map' for this edited volume is as follows: In Chapter 2 (Boon & Pringle), we set the philosophical context for the book and provide background information. Chapter 3 (Pringle & Withrington) sets out examples of freshwater conservation in action, contrasting approaches in the USA and the UK. Chapter 4 (Higgins & Duigan) discusses ways of evaluating the 'best' rivers and lakes for conservation, while Chapter 5 (Frissell & Bean) focuses on the importance of conservation evaluation when responding to catchment development. Chapter 6 (Langford & Frissell) goes a step further and looks at evaluating the potential for freshwater restoration, while Chapter 7 (Boon & Freeman) and Chapter 8 (Duker and Palmer) describe some of the techniques used in the conservation evaluation of rivers and lakes, respectively. Although the primary aim of the book is to compare approaches to freshwater evaluation in the UK and the USA, we have attempted to provide a flavour of what is happening in other places too. Towards that end, overviews of freshwater evaluation and conservation in Sweden (Chapter 9; Willen), Australia (Chapter 10; Nevill & Boulton) and South Africa (Chapter 11; O'Keeffe & Thirion) have been included in three short supplementary chapters, as examples of other developed countries where important contributions have been made to designing freshwater conservation assessment techniques. Chapter 12 (Abell & Bryer) examines this subject from a developing country perspective, where economic and social conditions impose a rather different conservation ethic from that of the developed world. Finally, the book ends with concluding remarks (Chapter 13; Pringle & Boon).

The field of freshwater conservation is not static. It constantly changes to meet the demands of new legislation, new environmental pressures, new understanding in science and the constantly shifting patterns in biological populations and communities. We hope that this book will contribute to the practice of conservation as it continues to develop in the UK, the USA and throughout the world.

References

Boon, P. J., Calow, P. & Petts, G. E. (eds.) (1992). *River Conservation and Management.* Chichester: John Wiley.

Boon, P. J., Davies, B. R. & Petts, G. E. (eds.) (2000). *Global Perspectives on River Conservation: Science, Policy and Practice.* Chichester: John Wiley.

Crooks, K. and Sanjayan, M. (eds.) (2006). *Connectivity Conservation.* Cambridge: Cambridge University Press.

Lawler, J. J., Aukema, J. E., Grant, J. B. *et al.* (2006). Conservation science: a 20-year report card. *Frontiers in Ecology and the Environment*, **4**, 473–80.

National Research Council (2005). *Valuing Ecosystem Services: Toward Better Environmental Decision-Making.* Committee on assessing and valuing the services of aquatic and related terrestrial ecosystems, Water Science and Technology Board. Washington, DC: National Academies Press.

Pringle, C. M. (2006). Hydrologic connectivity: a neglected dimension of conservation biology. In *Connectivity Conservation*, eds. K. Crooks & M. Sanjayan. Cambridge: Cambridge University Press, pp. 233–4.

Wishart, M. J., Davies, B. R., Boon, P. J. & Pringle, C. M. (2000). Global disparities in river conservation: 'First World' values and 'Third World' realities. In *Global Perspectives on River Conservation: Science, Policy and Practice*, eds. P. J. Boon, B. R. Davies & G. E. Petts. Chichester: John Wiley, pp. 354–69.

2 · *Background, philosophy and context*

PHILIP J. BOON AND CATHERINE M. PRINGLE

Introduction

Conservation for natural values, as an acceptable goal of freshwater resource management, is a relatively recent phenomenon in the USA and the UK, reflecting changes in societal needs and perceptions. These two countries are not alone in this, representing a much wider environmental awareness in the developed, and parts of the developing, world.

In the USA, the development of ideas regarding conservation in general, which influenced the emergence of freshwater conservation as a competing use, can be linked to three general philosophical conservation movements (Calicott, 1990): (1) the Romantic-Transcendental Conservation Ethic of the mid-1800s which stressed that nature has uses other than human economic gain (this was a basis for initial activism by many private conservation organizations whose goals were to save natural areas in pristine state for their inherent value); (2) the Resource Conservation Ethic at the turn of the twentieth century which is based on a utilitarian philosophy (whereby nature is equivalent to natural resources) leading to the 'multiple use concept' which is the current mandate of several US public land management agencies (e.g. US Forest Service and Bureau of Land Management); and (3) the Evolutionary-Ecological Land Ethic of the twentieth century based on the development of the science of ecology and evolution, where nature is viewed as a complicated and integrated system of interdependent processes and components. In the USA, NGOs have been quick to adopt this ethic – with federal and state agencies also including it in their new mandates.

While the Evolutionary-Ecological Land Ethic is the most biologically comprehensive approach to conservation, it is still only part of decision making since economic and social needs of people must also be met. However, it is clear that the science of ecology and evolution (as evinced by the Evolutionary-Ecological Land Ethic) is playing an increasingly greater role in the conservation of freshwater systems in both the USA and the UK.

Assessing the Conservation Value of Fresh Waters, ed. Philip J. Boon and Catherine M. Pringle. Published by Cambridge University Press.

Although the formal practice of nature conservation in the UK is little more than 50 years old (the term 'nature conservation' came to prominence only in the 1940s (Sheail, 1998)), its roots can be traced back for several millennia. The expansion of primitive agriculture in the Neolithic era (4500–2000 BC), the development of permanent settlements and the clearance of native woodland represent some of the key stages in the evolution of the British countryside closely linked with changing attitudes to the care and stewardship of the land.

By the late seventeenth century, 50% of England and Wales was committed to agriculture. Yet as people became isolated from the countryside in urban areas in the seventeenth and eighteenth centuries, negative attitudes regarding the value of natural wilderness began to change and a new awareness of wildlife and an enjoyment of natural scenery for its own sake began to grow (Allen, 1976; NCC, 1984; Evans, 1992). The eighteenth century may be considered the turning point towards modern conservation, with a move away from landscape architecture to an appreciation of more natural landforms (Nicholson, 1987). Two additional factors contributed to the development of the conservation movement in Britain. The first was the study of natural history that began in the seventeenth century as a middle-class social pastime (Nicholson, 1970; Evans, 1992) and flourished in the eighteenth, nineteenth and early twentieth centuries. The second was the formation of conservation member societies in the late nineteenth and early twentieth centuries, such as the Society for Protection of Birds (1891). Similar moves were afoot in the USA, with the creation of bodies such as the National Audubon Society (1902) and the National Wildlife Federation (1930) (Nicholson, 1987).

It is difficult to disentangle the development, specifically, of freshwater conservation from this more general evolution of nature conservation in Britain. Perhaps the greatest influence was the growth of freshwater science in the early twentieth century as a discipline in its own right, and in particular the founding of the Freshwater Biological Association in 1929.

In the USA, rivers have been central to the overall conservation movement (Palmer, 1994). For example, in 1910 the Hetch Hetchy Valley dam controversy on California's Tuolumne River was an important event in sparking public awareness of the importance of conservation. Although the battle to stop the dam was lost and this spectacular valley was flooded, the event gave rise to the National Park Service and the establishment of parks to protect wilderness and scenic areas. Only a few dams were stopped during the first half of the century – all in national parks. Aggressive opposition to dams did not occur until the 1960s and 1970s – with the founding event of the Echo Park Dam battle on the Green and Yampa Rivers in Colorado, which provided the impetus for the Wilderness Act in 1964 (Palmer, 1994). It is also instructive to note that one of the leading modern-day conservation NGOs in the USA, the Sierra Club, evolved out of the fight to protect the Grand Canyon from dam operations.

In Britain, much of the freshwater research undertaken in the first half of the twentieth century was strictly pure science; any attempts at applying scientific knowledge to freshwater conservation and management were largely restricted to tackling the legacy of the industrial and agricultural revolutions, especially the impacts of point-source pollution (Macan, 1951). There are parallels in the USA, where national attention was focused on cleaning up the most severe point-source pollution problems with the passing of the Clean Water Act of 1972. In Britain, by the time the NCC produced its first annual report in the mid-1970s, it was recognized that the problems confronting freshwater conservation were far wider than point-source pollution, including water abstraction, diffuse pollution, land drainage activities, river engineering, reservoir construction and water transfer schemes (NCC, 1975).

Likewise, the USA shifted from a focus on point-source pollution in the 1970s and 1980s to addressing also the regulation of diffuse pollutant inputs, maintaining critical instream flows and regulating river diversions and groundwater extraction – activities that have characterized the 1990s up to the present.

The importance of applying freshwater science to conservation will be discussed later in this chapter, and further comparisons of how conservation practices are implemented in the USA and the UK will be reviewed in Chapter 3.

The aim of this chapter is to set the context for the remainder of the book: first, to provide some brief background information on water resource legislation and agency structure in both the USA and the UK (for subsequent discussion in this chapter and others that follow); second, by looking briefly at the rationale underlying freshwater conservation; and third, to review opinions on which characteristics of rivers and lakes confer 'value' for conservation.

Legislation and agency structure in the USA and the UK

Clearly, socio-political and economic factors have helped shape approaches to conservation in both countries and are reflected by differences in legislation and the structure and mission of governmental and non-governmental agencies.

Legislative instruments relevant to freshwater conservation

The following sections summarize the principal legislative instruments relevant to freshwater conservation in the UK and the USA, including some that apply at a global or European level.

Ramsar Convention

The 'Convention on Wetlands of International Importance especially as Waterfowl Habitat' was signed in Ramsar, Iran, in 1971, and came into force in 1976. As of October 2008 there were 158 Contracting Parties with 1801 Ramsar sites covering 163×10^6 ha. The primary aim of the Convention is 'to stem the progressive encroachment on and loss of wetlands now and in the future' (www.jncc.gov.uk/legislation/conventions/ramsar.htm; www.ramsar.org). As its name implies, its focus is on water birds, although the criteria for selecting Ramsar sites now include other aspects of wetland conservation. The definition of a 'wetland' is a broad one: 'For the purpose of this Convention wetlands are areas of marsh, fen, peatland or water, whether natural or artificial, permanent or temporary, with water that is static or flowing, fresh, brackish or salt, including areas of marine water the depth of which at low tide does not exceed six metres', and that wetlands 'may incorporate riparian and coastal zones adjacent to the wetlands, and islands or bodies of marine water deeper than six metres at low tide lying within the wetlands'.

There are eight criteria for evaluating wetlands for designation as Ramsar sites. These state that a wetland should be considered internationally important if it

- contains a representative, rare or unique example of a natural or near natural wetland type found within the appropriate biogeographic region;
- supports vulnerable, endangered or critically endangered species or threatened ecological communities;
- supports populations of plant and/or animal species important for maintaining the biological diversity of a particular biogeographic region;
- supports plant and/or animal species at a critical stage in their life cycles, or provides refuge during adverse conditions;
- regularly supports 20 000 or more water birds;
- supports 1% of the individuals in a population of one species of water bird;
- supports a significant proportion of indigenous fish that contribute to global biological diversity;
- is an important source of food for fishes, or a spawning ground, nursery and/or migration path that fish stocks depend on. (www.ramsar.org/key_criteria.htm, 2004).

Convention on Biological Diversity

The Convention on Biological Diversity (CBD) was adopted at the Earth Summit in Rio de Janeiro, in June 1992 (www.biodiv.org). The objectives of the Convention are 'the conservation of biological diversity, the sustainable use of its components and the fair and equitable sharing of the benefits arising out of the utilization of genetic resources'. In addition, the signatories committed themselves to achieve by 2010 a significant reduction in the current rate of biodiversity loss at a global, regional and national level.

The CBD was initially signed by 157 governments, including the UK, with the USA signing a year later. At the time of writing (September 2007), 189 countries are party to the convention. As a demonstration of its commitment, the UK government rapidly responded by developing the UK Biodiversity Action Plan (BAP) (www.ukbap.org.uk/). This contains lists of priority species and habitats with target-based action plans. Many Local BAPs (LBAPs) have also been developed and implemented, and a computerized National Biodiversity Network (NBN) has been set up. The original UK BAP has recently been reviewed and in 2007 a revised list of 1149 priority species (double the original number) and 65 priority terrestrial, freshwater and marine habitats were identified. The current freshwater priority habitats are oligotrophic and dystrophic lakes, mesotrophic lakes, eutrophic standing waters, aquifer-fed naturally fluctuating water bodies, ponds and rivers. The CBD is given statutory backing in the UK through the Natural Environment and Rural Communities Act 2006 and the Nature Conservation (Scotland) Act 2004.

EC Habitats and Birds Directives

To implement the Bern Convention in Europe, the European Community adopted Council Directive 79/409/EEC on the Conservation of Wild Birds (the EC Birds Directive) in 1979, and Council Directive 92/43/EEC on the Conservation of Natural Habitats and of Wild Fauna and Flora (the EC Habitats Directive) in 1992. The Birds Directive provides a framework for the conservation and management of wild birds in Europe, while the objective of the Habitats Directive is 'to contribute towards ensuring biodiversity through the conservation of natural habitats and of wild fauna and flora in the European territory of the Member States to which the Treaty applies' (Article 2). One of the principal means of achieving these aims is by establishing a coherent European network of protected areas, designed to maintain the distribution and abundance of threatened species and habitats. This network (Natura 2000) comprises Special Areas of Conservation (SACs) designated under the Habitats Directive, and Special Protection

Areas (SPAs) designated under the EC Birds Directive. SACs can be selected for protecting specific, named habitat types listed on Annex I or species listed on Annex II of the Habitats Directive. SPAs are selected for nationally important numbers of birds listed on Annex I of the Birds Directive. The Birds Directive also embraces the Ramsar criteria for waterfowl.

Each Member State is required to contribute to the Natura 2000 series in proportion to the representation within its territory of these habitat types and species. The European Commission maintains an overview of the process, has the power to approve the sites chosen by Member States, and may require further sites to be designated if the candidate SAC list appears inadequate.

Site assessment criteria for the Annex I habitats and Annex II species of the Habitats Directive are given in Table 2.1, and are rather similar to nature conservation criteria used in the UK such as 'typicality' and 'size' (see Chapters 7 and 8).

Freshwater species included in Annex II of the Habitats Directive are otter *Lutra lutra*, Atlantic salmon *Salmo salar* and a number of other primarily river-dwelling fish, great-crested newt *Triturus cristatus*, freshwater pearl mussel *Margaritifera margaritifera*, a ramshorn snail *Anisus vorticulus*, white-clawed crayfish *Austropotamobius pallipes*, southern damselfly *Coenagrion mercuriale*, floating water-plantain *Luronium natans* and slender naiad *Najas flexilis*.

In the UK, the Directive has been transposed into national laws by means of the Conservation (Natural Habitats, &c.) Regulations 1994

Table 2.1 *Habitats Directive: site assessment criteria for habitats and species*

Annex I Habitats	Annex II Species
The degree of representativity of the natural habitat type on the site	The size and density of the population of the species present on the site in relation to the populations present within the national territory
The area of the site covered by the natural habitat type in relation to the total area covered by that natural habitat type within national territory	The degree of conservation of the features of the habitat which are important for the species concerned and restoration possibilities
The degree of conservation of the structure and function of the natural habitat type and restoration possibilities	The degree of isolation of the population present on the site in relation to the natural range of the species
A global assessment of the value of the site for conservation of the natural habitat concerned	A global assessment of the value of the site for conservation of the species concerned

(as amended), and the Conservation (Natural Habitats, &c.) Regulations (Northern Ireland) 1995 (as amended). These are known as 'the Habitats Regulations'. Most SACs on land or freshwater areas are underpinned by notification as Sites of Special Scientific Interest (SSSIs) (or as Areas of Special Scientific Interest (ASSIs) in Northern Ireland). In the case of SACs that are not notified as SSSIs, positive management is promoted by wider countryside measures, while protection relies on the provisions of the Habitats Regulations.

Water Framework Directive

The Water Framework Directive (WFD), adopted in 2000, is the single most significant piece of legislation covering aquatic environments ever enacted in Europe. The WFD takes a holistic approach to water management, applying to all surface waters (rivers, lakes, estuaries, coastal waters) as well as to groundwater. It aims to prevent further deterioration of aquatic ecosystems and to protect and enhance their status, and should have a profound effect on the ecological quality of Europe's waters. At the heart of the Directive is a requirement to produce river basin management plans, with the aim of achieving 'good surface water status' by 2015 (Boon & Lee, 2005). The emphasis on water quality alone, which has been prevalent in the past, has been abandoned for a wider definition of quality related to reference (near-natural) condition which has similarities to the 'ecosystem functionality' approach currently being promulgated in the USA. Assessments of 'ecological status' are closely linked to the effects of specific human pressures and impacts on phytoplankton, macrophytes, phytobenthos, benthic invertebrates and fish, together with physico-chemical and hydromorphological characteristics. The Directive requires each Member State to produce a typology for their surface waters and to classify the ecological status of individual water bodies on a five-point scale: high (at or very close to reference), good, moderate, poor and bad. Member States are required to take steps to protect high and good status water bodies and to return moderate, poor and bad water bodies to good ecological status.

In Scotland, the decision was taken to use primary legislation to transpose the requirements of the Directive, allowing full Scottish Parliamentary scrutiny. This resulted in the Water Environment and Water Services Act receiving Royal Assent in March 2003.

UK nature conservation legislation

The principal statutory mechanism for nature conservation (including freshwater conservation) in the UK was introduced in the Wildlife and

Countryside Act in 1981. This Act – which applies to England, Scotland and Wales, but not to Northern Ireland – has since been extensively modified by new legislation in England and Wales (the Countryside and Rights of Way Act 2000) and in Scotland (the Nature Conservation (Scotland) Act 2004). Separate, parallel legislation applies in Northern Ireland. The Act contains 'schedules' conferring strict protection for those species listed (including some freshwater ones such as freshwater pearl mussel *Margaritifera margaritifera* and vendace *Coregonus albula*) and prohibiting releasing into the wild of species considered to pose a threat to native habitats and biota (e.g. American signal crayfish *Pacifastacus leniusculus*).

This legislation also places a duty on the statutory conservation agencies – Natural England (NE), Scottish Natural Heritage (SNH) and the Countryside Council for Wales (CCW) – to select areas of land or water containing plants, animals, geological or geomorphological features of special interest, and to notify them to land owners and occupiers, to the appropriate government minister, to the local planning authority and to a range of public bodies. Owners and occupiers receive a description of the features for which the site is notified, a map of the site boundary, and a list of activities that are likely to cause damage and for which consent is required before they are carried out.

In 1989, the NCC (the forerunner in Britain to NE, SNH and CCW) published guidelines for selecting biological SSSIs (NCC, 1989). These were based on generally accepted conservation criteria, first set out in detail in *A Nature Conservation Review* (Ratcliffe, 1977). Although many things have changed since the guidelines were published, they remain in use and have yet to be revised.

Under the Nature Conservation (Scotland) Act 2004, the Natural Environment and Rural Communities Act 2006 and parallel legislation for Northern Ireland, public authorities must have regard for the conservation of biodiversity. Lists of habitats and species of principal importance for the conservation of biodiversity must be published by the governments of England, Wales, Scotland and Northern Ireland and the administrations must take steps to further the conservation of these habitats and species.

The US Wild and Scenic Rivers Act

The Wild and Scenic Rivers Act (passed in 1968) declared that it would be the policy of the United States to protect certain rivers of the Nation that possessed 'outstandingly remarkable' scenic, recreational, historic, cultural or natural values including geological, fish and wildlife features. Protection entailed preserving these outstanding rivers or river segments in their free-flowing condition. The Act designated 156 river segments to be included in the wild and scenic rivers system and specified a process by which additional rivers could be added to the system 'from time to time'. The primary criterion for inclusion in the system was that the river or river section was free-flowing. For designation as a 'Wild river area', the Act specified that the catchment or shorelines should be 'essentially primitive', access limited to trails and the waters unpolluted. A 'Scenic river area' could be accessible in places by roads, and a 'Recreational river area' could have some development along the river and ready access by road or rail. The Act was a political mechanism for protecting sections of certain rivers valued by the public for their 'wild', scenic or recreational qualities from damming and development that would directly impair water quality or scenic values.

The US Endangered Species Act

The Endangered Species Act (passed in 1973) combined and considerably strengthened the provisions of its predecessors – the Endangered Species Preservation Act 1966, and the Endangered Species Conservation Act 1969 (www.fws.gov/endangered/esasum.html). Its provisions included:

- US and foreign species lists were combined, with uniform provisions applied to both.
- Categories of 'endangered' and 'threatened' were defined (section 3).
- Plants and all classes of invertebrates were eligible for protection, as they are under the Convention on International Trade in Endangered Species of Wild Fauna and Flora (CITES).
- All federal agencies were required to undertake programmes for the conservation of endangered and threatened species, and were prohibited from authorizing, funding or carrying out any action that would jeopardize a listed species or destroy or modify its 'critical habitat'.
- Authority was provided to acquire land for listed animals and for plants listed under CITES.

Significant amendments were enacted in 1978, 1982 and 1988, while the overall framework of the 1973 Act has remained essentially unchanged.

The US Clean Water Act

The Clean Water Act (CWA) and subsequent rule-making provides a mechanism, the 'Outstanding Waters' designation, for identifying the highest quality streams and rivers, again based on perceived aesthetic, environmental and ecological value to the public. The purpose of the CWA is to 'restore and maintain the chemical, physical, and biological integrity of the Nation's waters', with the goal that all waters should be 'fishable and swimmable'. To accomplish this, the Act and subsequent rule-making has required that states designate 'uses' for all water bodies, and water quality standards necessary to protect those designated uses. Furthermore, the states are required to develop and implement measures (1) to protect existing uses from degradation, (2) to minimize any lowering of quality in water bodies that surpass clean water standards ('high-quality waters') and (3) to protect 'Outstanding Waters'.

Organizations relevant to freshwater conservation

It is not our goal here to create a comprehensive list of all bodies in the US and the UK with some role in aquatic resource management and conservation. The information we present serves simply to illustrate that there are large differences between the organizational structures in both countries. Figures 2.1 and 2.2 provide annotated illustrations of these organizational differences in the USA and the UK.

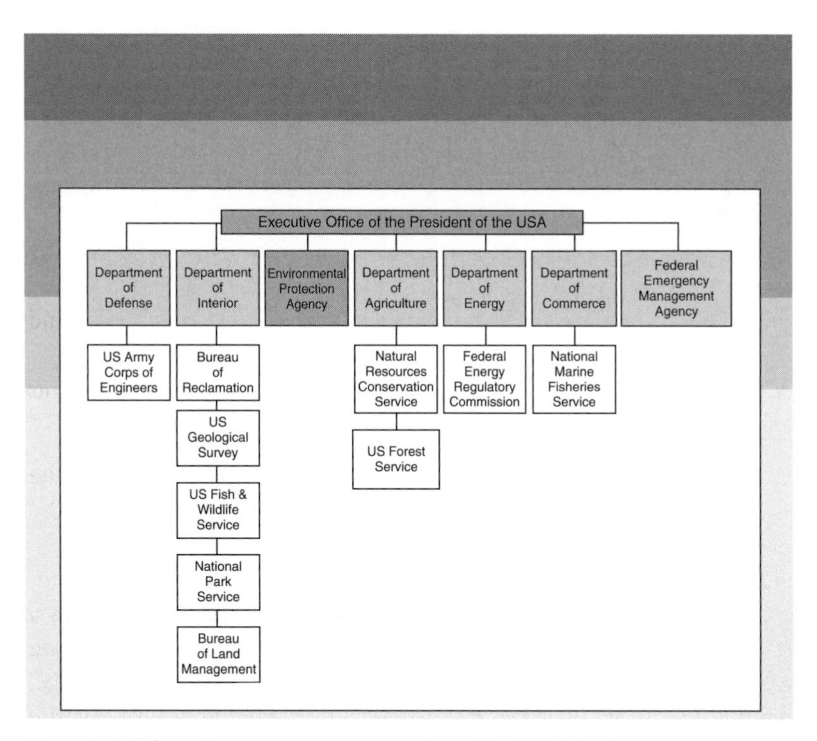

Figure 2.1. Selected water resources agencies of the US federal government.

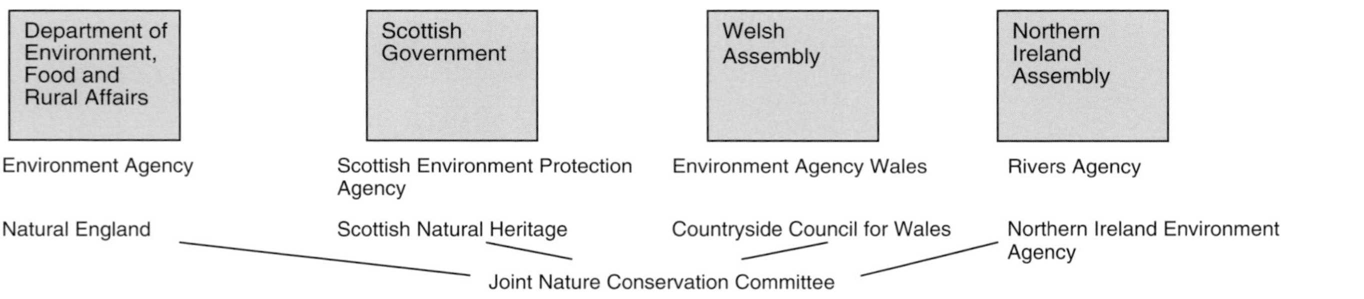

Organization	Primary role
Environment Agency (England and Wales)	Government agency with duties that include regulating pollution, water abstraction and impoundment; managing inland fisheries; supervising and implementing flood defences; managing navigation on rivers; freshwater monitoring. EA also has a duty to promote nature conservation.
Scottish Environment Protection Agency	Government agency in Scotland with similar duties to EA, except for fisheries and flood defence.
Joint Nature Conservation Committee	Government agency, funded by the statutory countryside agencies (NE, SNH, CCW, NIEA), providing a UK overview of nature conservation.
Natural England (NE), Scottish Natural Heritage (SNH), Countryside Council for Wales (CCW)	Government agencies in Great Britain responsible for nature conservation, countryside access, recreation and enjoyment.
Northern Ireland Environment Agency	Part of the government's Department of the Environment in Northern Ireland, with a duty to protect, conserve and promote the natural environment and built heritage.
Rivers Agency	An Executive Agency of the Department of Agriculture and Rural Development. It is the statutory drainage and flood defence authority for Northern Ireland.

Figure 2.2. The main organizations with direct or indirect responsibilities for freshwater management and conservation in the UK.

Other relevant organizations	Primary role
Water companies	Private companies in England and Wales providing water and/or wastewater treatment services for England and Wales.
Scottish Water	Publicly owned business, providing water and wastewater treatment services for Scotland.
Northern Ireland Water	A government-owned company, and the sole provider of water and sewerage services in Northern Ireland.
District Salmon Fisheries Boards	Bodies set up by statute, but funded by fishery proprietors, to protect, preserve and develop salmon fisheries in Scotland.
British Waterways	A public corporation responsible for maintaining 2000 miles (3220 km) of canals and navigable rivers in England, Scotland and Wales, primarily as a leisure resource.
Forestry Commission	The Government department responsible for forestry policy throughout Great Britain, licensing tree planting and felling, providing woodland grants, and managing state forests. Provides published guidance on the environmental effects of land-use, pollutant inputs and forest operations on fresh water.
Natural Environment Research Council	Government research body, comprising research institutes such as the Centre for Ecology and Hydrology.

Figure 2.2. (cont.)

Shifting philosophies: expanding freshwater resource management to include conservation and sustainability

Despite an increasing appreciation, both in the USA and the UK, of nature conservation as an underlying principle in resource management, there remains a wide range of opinions on the matter.

In the USA, the philosophy within many government agencies has shifted towards merging management objectives with nature conservation. This is evidenced by the changing mandate of many federal and state agencies that were traditionally involved in management of water resources for their utilitarian values. Over the last decade and a half, protection and restoration of natural values have been stressed through a wide range of legal duties authorized by the US Congress. Historically, many federal agencies were set up to promote the settlement of frontier territories. After the Second World War, the federal focus shifted from water development to environmental protection and conservation. New agencies were created, such as the US Environmental Protection Agency in 1970, while others were given expanded missions such as the US Fish and Wildlife Service (the only US federal agency with the primary responsibility to conserve and enhance fish, wildlife and plants) (see Chapter 3 for additional information).

In the UK, the activities that are usually considered to be part of freshwater management – such as regulating polluting discharges, abstraction control, flow and water level manipulation and river channel maintenance – have frequently been viewed in a rather different way from the practice of nature conservation (Boon, 1994; Howell, 1994). Furthermore, nature conservation, rather than being perceived as fundamental to sustainable environmental management, was often seen as little more than another competing 'use' alongside fisheries or navigation. This approach seems to have been accepted even by the Nature Conservancy (then Britain's statutory conservation body) more than 30 years ago.

> In practice, the element of competition creates a situation in which the claims of nature conservation ... have to be weighed against other forms of use in the overall national interest. This means that while the nature conservation interests in general would wish for as much as possible of the total remaining wildlife resource of Britain to be safeguarded ... the Nature Conservancy, as a responsible Government body, has to accept that compromise is often necessary (Nature Conservancy, 1972).

Boon and Howell (1997) explored the concept of freshwater nature conservation as a use in more depth, by means of a questionnaire to staff in UK regulatory and nature conservation bodies, and academic and research institutions. Opinion was divided, with 47% of respondents stating that nature conservation should be defined as a use, compared with 41% who felt that it should not. Currently, government agencies in both the UK and the USA, and those with an international remit, are keen to stress that people are an inseparable part of nature, and that human uses of ecosystems do not necessarily need to compromise natural values. Within the UK, the mission statements of the statutory conservation agencies – NE (www.naturalengland.org.uk), SNH (www.snh.org.uk), CCW (www.ccw.gov.uk) and the Northern Ireland Environment Agency (www.ni-environment.gov.uk) – and of voluntary bodies such as the Wildfowl and Wetlands Trust (www.wwt.org.uk), all place similar emphasis on sustainable management of land and water. Indeed, the Natural Heritage (Scotland) Act 1991, which founded SNH, introduced the word 'sustainable' into UK legislation for the first time.

Non-governmental conservation organizations in the USA (i.e. voluntary conservation bodies) have largely embraced the idea of sustainability as a guiding objective (see Chapter 3). For example, the largest voluntary conservation body in the USA is The Nature Conservancy, working in all 50 states (and in 27 countries). Its philosophical position on freshwater conservation is very similar to that of the international and UK organizations quoted above:

> There is growing recognition that healthy freshwater ecosystems provide valuable natural services – such as water purification, plant and animal foods, flood control, recreation, nutrient cycling, and biodiversity maintenance – that are being lost to improper water management. The Nature Conservancy believes there is a way to find a balance between the freshwater needs of people and ecosystems. The Conservancy's response to this challenge is to advocate Ecologically Sustainable Water Management (ESWM) – the compatible integration of human and natural ecosystem needs (www.nature.org/initiatives/freshwater/about/).

International perspectives

Several international conservation bodies are at the forefront of promoting the sustainability ethic. Indeed, the International Union for Conservation of Nature and Natural Resources (IUCN), an association of government institutions and NGOs, changed its name to accommodate this view as early as 1956, just 8 years after it was first established (Christoffersen, 1997). The aim of IUCN's founders was to encourage international efforts in protecting habitats and species against damaging human activities; hence its original name – the International Union for the Protection of Nature. However, the name was changed when the leadership realized that protected areas and threatened species could be protected more effectively if local people considered it in their own interest to do so (Christoffersen, 1997). As ever, there is a balance to be struck. In emphasizing that freshwater ecosystems provide 'goods' and 'services' to society, there is a danger that the inherent 'rights' of natural habitats and species to exist may be ignored. IUCN, for instance, summarizes the species diversity of 'wetlands' (a term which it uses to include rivers and lakes) as follows:

> This biological diversity is attractive as well as useful to humanity. Wetlands sustain thousands of communities with a wide range of products, such as drinking water, fish, food and timber ... Wetlands protect us against floods and droughts and offer space for recreation, and they may have strong cultural or religious values ... Ecosystems provide an estimated US $33 trillion per year to societies, of which an estimated 26% comes from freshwater ecosystems (www.iucn.org/themes/wetlands/wetlands.html).

Another leading international NGO – World Wildlife Fund (WWF) – emphasizes the people and nature approach in its work on Integrated Water Resource Management (IWRM):

> WWF believes that ... social and development goals are not in conflict with biodiversity conservation. Instead, Integrated Water Resource Management can contribute to poverty reduction and sustainable development. WWF's framework for achieving these goals includes tools of its Integrated River Basin Management (IRBM), an approach that conserves rivers from source to sea (www.panda.org/about_wwf/what_we_do/freshwater/our_solutions/rivers/iwrm.cfm).

The emphasis on IWRM as a means of bringing together nature conservation and sustainable development emerges time and again from conservation NGOs. In April 2005, six of the largest global NGOs – IUCN, WWF, Conservation International, Wetlands International, Bird Life International, and The Nature Conservancy – wrote a joint statement for consideration at the 13th Session of the Commission on Sustainable Development: 'We see a clear and urgent need for integration of ecosystem conservation into water and sanitation policies and practices including through poverty reduction strategies, national sustainable development strategies and national budgets and investment plans' (www.panda.org/downloads/freshwater/csd13ceoletter.pdf).

The evidence presented above demonstrates that the conceptual barriers between freshwater 'conservation' and freshwater 'management' have begun to erode in both countries, but to differing degrees. Nevertheless, Boon (2000), writing specifically about rivers, argued that conservation and management perspectives are different, despite many areas of overlap and a set of shared goals. A freshwater conservation perspective highlights the special (the natural, the rare, the threatened, the diverse) yet still appreciates the ordinary. It stresses 'non-use' (e.g. aesthetic) values while realizing the importance of 'use' (e.g. economic) values. The freshwater management perspective focuses on the ordinary but does not underestimate the value of the uncommon. It emphasizes uses, but does not neglect non-use values. Both perspectives are needed, and assessments of nature conservation value are therefore an essential element in the wider arena of IWRM.

The role of scientific societies and the scientific community in promoting freshwater conservation

Methods for assessing freshwater conservation status should be grounded in freshwater science. Judgements on the value of freshwater ecosystems, and consequential decisions on their management, require an understanding of the physical, chemical and biological processes critical to maintaining freshwater habitats and species. Although it seems obvious that scientists must be involved in conservation, the evidence from both sides of the Atlantic suggests that there is a long way to go in making this work effectively.

In this context two particular questions are worth exploring: (1) To what extent are freshwater scientists in universities and research institutes involved in work relating to freshwater conservation? (2) Do professional societies help to fulfil a role in promoting the application of science to conservation?

Scientific societies have considerable potential for promoting the integration of science and freshwater conservation through publication, informal exchange of information, and sponsorship of conferences and symposia (Pringle *et al.*, 1993). Some, such as the International Association of Limnology (SIL), operate at a global level. Others, while having an international membership, are predominantly associated with one country.

In the USA, the American Fisheries Society (AFS) is a good example of effective involvement of a scientific society in freshwater conservation issues. The AFS has been involved in key environmental issues since its founding in 1870. The Society was founded with the dual purpose of strengthening the science of fish biology and ecology and advocating the conservation of fisheries resources. This explains, in part, how the AFS has avoided the polarity that develops in many scientific societies around issues such as science versus management and basic versus applied (Dewberry & Pringle, 1994). One of AFS's first activities in the late 1800s was to lodge protests with the US State Department and with Canada against obstructions in the St Lawrence River that restricted the migration of salmon (Benson, 1970). The AFS is involved in lobbying and advocacy activities, has supported agency fishery programme budgets, developed legislation and commented on legislative initiatives of others. Other key activities include forming a coalition of fishermen to conserve the habitat, sponsoring of the Mississippi Inter-jurisdictional Cooperative Resources Agreement and launching of the Fisheries Action Network (Brouha, 1993).

In the UK, the Freshwater Biological Association (FBA) provides a very different example of the role of scientific societies in freshwater conservation. The FBA is a Registered Charity with 71% of its 1681 members from the UK. Founded in 1929, it aims to promote freshwater science through innovative research, scientific meetings and publications, and by providing sound, independent opinion. From its inception it was the principal focus of freshwater research in the UK, with many of its staff (e.g. T. T. Macan, E. B. Worthington, W. H. Pearsall, J. F. Talling) pivotal figures in the worldwide development of freshwater science. Now, much of the FBA's research activity has been transferred to the government-funded Centre for Ecology and Hydrology (CEH). However, the publications for which the FBA has become famous (such as its taxonomic keys) are still an important part of its work, and its library remains one of the finest collections of freshwater publications in the world. Over the past 5–10 years the FBA has begun to focus rather more on applied issues relevant to freshwater management and conservation, through publications such as *Lake Assessment and the EC Water Framework Directive* (FBA, 2001a) and *European Temporary Ponds* (FBA, 2001b), and through its annual scientific meetings such as the most recent one on 'Restoration and Recovery of Fresh Waters'. Unlike AFS in the USA, the FBA is not a lobbying organization; any involvement in campaigning, advocacy or advice is generally by individual members in their own right, rather than on behalf of the FBA.

In the USA, over the last two decades other scientific societies such as the North American Benthological Society (NABS) in the USA (73% US membership and 7% Canadian) have become involved in freshwater conservation issues to a greater extent. For example, conservation became a prominent part of the NABS scientific programme at the NABS meeting in 1992, which included a plenary symposium on *Current Issues in Freshwater Conservation* (Pringle & Aumen, 1993; Pringle *et al.*, 1993) This symposium illustrates well the key role that NGOs play in freshwater conservation in the USA, as it was organized around presentations by seven major conservation NGOs and one professional society (AFS): American Rivers Inc., Pacific Rivers Council, Wilderness Society, The Nature Conservancy, Sierra Club, Trout Unlimited, Tatshenshini Wild, and the American Fisheries Society. One of the activities resulting from this symposium was the revitalization of the NABS Conservation and Environmental Issues Committee which is currently very active in addressing a variety of freshwater conservation issues. Over the last decade, NABS has increasingly featured conservation-related symposia and plenary sessions.

In response to the US national need for a more predictive understanding of how freshwater ecosystems respond to environmental change, a large number of concerned aquatic ecologists and limnologists came together to develop a research agenda in the early 1990s (the Freshwater Imperative (Naiman *et al.*, 1995)). This represents a two-year effort and identifies critical research needs that will help to reverse the accelerating deterioration of the biological integrity of water resources. The venture was sponsored by several federal agencies (including the US EPA) for the purposes of developing a scientific basis for sound management decisions associated with the nation's freshwater resources.

It is difficult to find a similar parallel to this in the UK, where freshwater conservation and management tends to be focused on those working in the statutory environment and conservation agencies, with less involvement by academic scientists. It also highlights the contrasting fortunes of freshwater science in the USA and the UK: whereas freshwater science continues to be relatively healthy in the USA, its gradual demise in the UK is a source of great concern, and a potential threat to the future of freshwater conservation efforts. A recent report (www.fba.org.uk/pdf/ReviewOfFreshwaterEcology.pdf) produced by a consortium of freshwater scientists makes depressing reading. The last 20 years has seen large reductions in the numbers of staff employed by bodies such as CEH, as well as in the number of academic freshwater research scientists. For example, the report states that from 1981 to 1986, 59 individuals in the UK universities were regularly publishing scientific papers in major freshwater biological journals. In 1999–2004 this number had dropped to 32.

In the USA, scientific and technological advice that can be used in management and policy decisions is provided to the federal government via the US National Academy of Sciences (NAS). The NAS is a society of distinguished scholars engaged in scientific and engineering research with a mandate that requires it to advise the federal government on scientific and technical matters. The National Research Council (NRC) is the principal operating agency of both the NAS and the NAE (National Academy of Engineering), which is an organization of outstanding engineers that is parallel to the NAS. The NRC brings together committees of scientists, often from several different disciplines, to address scientific issues and environmental challenges. The deliberations, research activities, and opinions of these committees are synthesized into published volumes.

Over the last two decades, many of these committees have been established (often at the request of a number of federal agencies) to address water resource issues. Focal areas include the restoration of aquatic ecosystems (NRC, 1992), riparian zone management (NRC, 2002a), valuing ecosystem services with an emphasis on aquatic and related terrestrial ecosystems (NRC, 2005), and establishing a future water resources agenda (NRC, 2001a). Some committees target environmental challenges associated with specific freshwater ecosystems (NRC, 2002b, 2004), while others focus on management issues such as catchment management for potable water supplies (NRC, 2000) or assessment of new water quality management approaches such as the total maximum daily load (TMDL) (NRC, 2001b).

While the UK has no analogue to the US National Research Council, the government clearly recognizes the importance of science and its role in underpinning, for example, the EC WFD. The government also recognizes that any major investment in improving freshwater environments must be supported by rigorous and defensible scientific assessments. Consequently, it has set up a UK Technical Advisory Group (UKTAG: www.wfduk.org) and a series of subgroups ('Task Teams') comprising scientific and technical staff from the statutory environment and conservation agencies. The work of these groups has included the development of new techniques for assessing and monitoring 'ecological status' under the WFD. For example, the present lack of any tools for recording and evaluating the morphological and hydrological features of lakes has led to a new method of 'Lake Habitat Survey' (Rowan *et al.*, 2006). This method will contribute to assessments of the 'condition' of designated conservation sites in the UK as well as assisting in monitoring them under the WFD. Much of this work is overseen by individual scientists in UKTAG or its participating agencies, and carried out on contract by universities, research institutes and consultancies. As scientific information is assembled, it feeds into the development of policy for implementing the Directive (e.g. for the environment agencies to use when regulating developments on rivers and lakes) (www.sepa.org.uk/wfd/regimes/index.htm).

As in the USA, government agencies in the UK commonly use available relevant scientific information in management. For example, the development of forestry in the UK is required to consider the impacts that forest planting and management have on freshwater environments. The Forestry Commission (FC) is a British government agency with a mission 'to protect and expand Britain's forests and woodlands and to increase their value to society and the environment'. The objective of the FC is to take the lead, on behalf of the government in England, Scotland and Wales, in the development and promotion of sustainable forest management and to support its achievement nationally. To help do this, the FC has published the *Forests and Water Guidelines*, which applies scientific understanding of the links between forestry activities and freshwater environments. The latest edition (Forestry Commission, 2003) sets out practical advice on how to mitigate the effects of planting, forest management and harvesting on siltation and turbidity, acidification, nutrient enrichment, chemical pollution, and flow regimes. The guidance is set within the context of catchment processes, and the ecological requirements of freshwater habitats and species.

Freshwater conservation is increasingly a topic in peer-reviewed scientific publications such as *Aquatic Conservation*, *Ecological Applications*, *Frontiers*, *American Journal of Water Resources*, *BioScience*, *Environmental Management* and *River Research and Applications*. Some of these journals are relatively new (e.g. *Frontiers*) and have been started within the last decade, providing an excellent outlet for freshwater conservation research.

Perceptions of nature conservation value

So far, this chapter has sought to set the context for the remainder of the book – in particular by demonstrating, first, that there is a need for methods of freshwater conservation evaluation (because traditional assessments of freshwater quality for resource management and regulation do not take into account all aspects of nature conservation); and second, that freshwater scientists must be involved in freshwater conservation and in the development of techniques for conservation assessment.

The third element that completes this contextual outline is to consider the natural values of rivers and lakes that are deemed to be important for conservation. It is not the role of environmental legislation to describe in detail how nature conservation evaluation is to be undertaken. Yet some of the criteria that are used for assessing the nature conservation values of rivers and lakes are implicit (and occasionally explicit) in the Acts, conventions and directives described earlier in this chapter. Unfortunately, the way in which legislation is interpreted and applied has often led to a lack of objectivity, rigour, and repeatability in conservation evaluation, and attempts have been made in both the UK and the USA to address these problems (see Chapters 7 and 8).

In preparing this chapter, we were especially interested in addressing three questions: (1) What characteristics of rivers and lakes do freshwater specialists believe to be important indicators of nature conservation value? (2) To what extent are these perceptions common to both the UK and the USA? (3) Are there any significant disparities in the views held by different groups of professionals?

Canvassing the views of others

In order to address these questions, a short questionnaire was assembled and distributed to selected individuals in the USA and the UK. Recipients were asked to assign a score to a range of pre-selected conservation criteria and attributes (Tables 2.2 and 2.3), and to propose any additional features that they felt may have been overlooked. The general categories of conservation attributes listed were derived from SERCON (Boon *et al.*,

Table 2.2 *Mean and median criterion scores for rivers, allocated by all respondents, and separately for respondents in the UK and the USA*

	Total			UK			USA		
Criteria	*N*	Mean	Median	*N*	Mean	Median	*N*	Mean	Median
Naturalness	61	8.00	7.88	33	7.73	7.63	28	8.32	8.19
Rarity	61	7.86	8.00	33	7.32	7.29	28	8.49	8.79
Diversity/richness	61	6.94	7.20	33	6.32	6.20	28	7.69	8.00
Representativeness	61	6.80	7.33	33	6.47	7.00	28	7.18	7.58
Special features	61	6.33	6.50	33	5.53	5.50	28	7.27	7.25

Table 2.3 *Mean and median criterion scores for lakes, allocated by all respondents, and separately for respondents in the UK and the USA*

	Total			UK			USA		
Criteria	*N*	Mean	Median	*N*	Mean	Median	*N*	Mean	Median
Naturalness	33	7.98	8.00	21	7.77	7.50	12	8.34	8.38
Rarity	33	7.91	8.14	21	7.57	7.71	12	8.50	8.71
Diversity/richness	33	7.24	7.20	21	6.66	7.00	12	8.25	8.40
Representativeness	33	6.64	6.83	21	6.24	6.50	12	7.33	7.58
Special features	33	6.11	7.00	21	5.55	5.50	12	7.08	7.75

1997, 2002) and from Dunn (2004) and the survey method closely followed that of Dunn (2004) for Australian rivers. Views were sought only on the relative values of freshwater habitats and species, and did not cover other aspects such as aesthetic or recreational value. Respondents were asked how important they considered each attribute to be in terms of its nature conservation value by assigning a score of 1–10, where 1 represented very low value and 10 very high value (with zero representing no value).

Statistical analysis

In total, 81 responses were received, 43 from the UK and 38 from the USA. Fewer responses covered lakes than rivers (45 compared with 74). Responses were also classified into three broad groups, under the headings: A: 'Universities, research institutes', B: 'Conservation bodies' and C: 'River, lake and land managers/ environmental regulators'.

As the distributional assumptions of parametric statistical tests were not readily satisfied, non-parametric tests were used to compare scoring between attributes and between groups of respondents. To compare each of the main criteria (naturalness, rarity, etc.) the mean of each group of attributes was calculated. The Wilcoxon signed-rank test for paired samples was then applied to look for evidence that one criterion mean tended to be higher or lower than another. The Mann – Whitney *U* test was used to examine whether UK respondents tended to give higher or lower scores to a particular attribute than US respondents. To compare the three groups – researchers, conservationists and resource managers – a Kruskal – Wallis one-way analysis of variance was used. Finally, variation between all attributes within each of the main criteria was examined with Friedman's test.

Results

Tables 2.2 and 2.3 give the overall conservation criterion scores for rivers and lakes, respectively. The highest mean scores in both categories were assigned to naturalness attributes, closely followed by rarity. The remaining criteria (for both rivers and lakes) were scored in the order diversity/richness, representativeness, and special features. Statistical pairwise comparisons of these criteria showed that mean scores for naturalness and rarity were not significantly different either for rivers or lakes, but that both were significantly different from the other three criteria ($p < 0.01$ for river comparisons, $p < 0.01$ or < 0.05 for lake comparisons).

In every case (both for rivers and lakes), the mean scores for the five criteria were higher from US than from UK respondents (Tables 2.2 and 2.3). For rivers, these differences were statistically significant ($p < 0.05$) for all except representativeness, while for lakes, differences for rarity, diversity/richness and special features were significant ($p < 0.05$) but those for naturalness and representativeness were not. When mean scores allocated for each individual attribute were analysed (Tables 2.4 and 2.5), every one that showed significant differences between US and UK responses were given higher scores by US respondents.

Analysis of the differences in scores allocated by the three sectoral groups (A, B, C) used data from the whole US/UK dataset because of its limited size. For Groups A and C, the top three criteria in score order were rarity, naturalness and diversity/richness (rivers and lakes); for Group B scoring rivers, the order was naturalness, rarity and representativeness, and, for lakes, naturalness, rarity and diversity/richness (Tables 2.6 and 2.7). Of the 56 attributes examined (28 for rivers and 28 for lakes), 11 were significantly different between groups ($p < 0.05$, Table 2.8). On every occasion, Group B respondents scored these attributes lower than Groups A and C.

Interpretation

Both the scores and the accompanying comments strongly endorse the view that naturalness is the key quality prized the most highly in freshwater ecosystems, in both the UK and the USA.

Of the four conservation criteria listed above, representativeness (= typicalness) was the one that generated the most uncertainty, leading to lower scores in most cases and with some respondents choosing not to allocate scores at all. However, several respondents stated that representativeness is needed so that assessments of diversity or richness can be set in context.

Table 2.4 *Mean and median attribute scores for rivers, allocated by all respondents, and separately for respondents in the UK and the USA*

Criteria and attributes	Total			UK			USA		
	N	Mean	Median	*N*	Mean	Median	*N*	Mean	Median
Naturalness									
Channel and banks	73	8.68*	9.00	37	8.38	8.00	36	9.00*	9.50
Riparian zone	73	8.07**	8.00	37	7.32	7.00	36	8.83***	9.00
Flow regime	73	8.48**	9.00	37	8.05	8.00	36	8.92**	10.00
Water chemistry	73	7.82	8.00	37	7.81	8.00	36	7.83	8.00
Ecological processes (e.g. energy flow, groundwater linkages)	73	8.00	8.00	37	7.92	8.00	36	8.08	8.00
Native freshwater plants (lacking aliens)	73	7.36	7.00	37	7.51	8.00	36	7.19	7.00
Native freshwater invertebrates	73	7.85	8.00	37	7.65	8.00	36	8.06	8.00
Native freshwater fish	73	7.96**	8.00	37	7.57	8.00	36	8.36**	9.00
Representativeness									
Freshwater plants representative of river type	67	6.28	7.00	34	6.35	7.00	33	6.21	7.00
Freshwater invertebrates representative of river type	67	6.67	7.00	34	6.47	7.00	33	6.88	7.00
Freshwater fish representative of river type	67	6.72*	7.00	34	6.21	6.00	33	7.24*	8.00
Physical features representative of river type	67	6.90*	7.00	34	6.41	7.00	33	7.39*	8.00
Ecological processes representative of river type	67	7.01	8.00	34	6.68	7.00	33	7.36	8.00
Overall representativeness of river ecosystem	67	6.88	7.00	34	6.62	7.00	33	7.15	8.00
Rarity									
Rare or threatened communities/ecosystems	67	8.15**	8.00	36	7.58	8.00	31	8.81**	10.00
Rare or threatened plant species	67	7.36*	8.00	36	6.89	7.00	31	7.90*	8.00
Rare or threatened invertebrate species	67	7.69**	8.00	36	7.00	7.00	31	8.48***	9.00
Rare or threatened fish species	67	7.81**	8.00	36	7.06	7.50	31	8.68***	9.00

Table 2.4 (*cont.*)

Criteria and attributes	Total			UK			USA		
	N	Mean	Median	*N*	Mean	Median	*N*	Mean	Median
Endemic species	67	8.25★★	8.00	36	7.58	8.00	31	9.03★★	9.00
Statutorily protected species	67	7.38★★	8.00	36	6.60	7.00	31	8.29★★★	9.00
Rare or threatened ecological, hydrological or geomorphological processes	67	8.16	8.00	36	7.86	8.00	31	8.52	9.00
Diversity/richness									
Freshwater plants	71	6.68	7.00	36	6.56	7.00	35	6.80	7.00
Freshwater invertebrates	71	7.08★	8.00	36	6.53	7.00	35	7.66★	8.00
Freshwater fish	71	7.04★★	7.00	36	6.19	6.00	35	7.91★★	8.00
Instream habitats	71	7.41★	8.00	36	6.92	7.00	35	7.91★	8.00
Adjacent habitats	71	6.69★	7.00	36	6.17	6.00	35	7.23★	8.00
Special features									
Important floodplains and/or adjacent wetland habitats (e.g. fen, marsh, wet woodland)	71	7.86★★	8.00	36	7.03	7.00	35	8.71★★★	9.00
Presence of economically important species (e.g. salmon)	71	5.07★★	5.00	36	4.17	4.00	35	6.00★★	6.00

★$p < 0.05$
★★$p < 0.01$
★★★$p < 0.001$

Table 2.5 *Mean and median attribute scores for lakes, allocated by all respondents, and separately for respondents in the UK and the USA*

	Total			UK			USA		
Criteria and attributes	*N*	Mean	Median	*N*	Mean	Median	*N*	Mean	Median
Naturalness									
Lake littoral and shoreline zones	42	8.50	8.00	25	8.20	8.00	17	8.94	10.00
Riparian zone	42	7.33**	7.50	25	6.60	7.00	17	8.41**	9.00
Hydrological regime (e.g. residence time, water level fluctuation)	42	7.98	8.00	25	7.56	8.00	17	8.59	8.00
Water chemistry	42	8.19	8.00	25	8.24	8.00	17	8.12	8.00
Ecological processes	42	8.07	8.00	25	8.20	8.00	17	7.88	8.00
Native freshwater plants	42	7.93	8.00	25	7.88	8.00	17	8.00	8.00
Native freshwater invertebrates	42	7.95	8.00	25	7.60	7.00	17	8.47	8.00
Native freshwater fish	42	8.00	8.00	25	7.92	8.00	17	8.12	8.00
Representativeness									
Freshwater plants representative of lake type	41	6.66	7.00	25	6.48	7.00	16	6.94	7.00
Freshwater invertebrates representative of lake type	41	6.71	7.00	25	6.32	7.00	16	7.31	7.50
Freshwater fish representative of lake type	41	6.90*	7.00	25	6.40	7.00	16	7.69*	8.00
Physical features representative of lake type	41	6.66*	7.00	25	6.12	6.00	16	7.50*	8.00
Ecological processes representative of lake type	41	6.98**	7.00	25	6.36	6.00	16	7.94**	8.00
Overall representativeness of lake ecosystem	41	6.98	7.00	25	6.48	7.00	16	7.75	8.00
Rarity									
Rare or threatened communities/ecosystems	40	8.38**	9.00	25	7.80	8.00	15	9.33**	10.00
Rare or threatened plant species	40	7.75*	8.00	25	7.24	8.00	15	8.60*	10.00
Rare or threatened invertebrate species	40	7.85**	8.00	25	7.24	8.00	15	8.87**	10.00
Rare or threatened fish species	40	7.70*	8.00	25	7.32	8.00	15	8.33*	9.00

Table 2.5 (*cont.*)

Criteria and attributes	Total			UK			USA		
	N	Mean	Median	*N*	Mean	Median	*N*	Mean	Median
Endemic species	40	8.48*	9.00	25	7.92	8.00	15	9.40*	10.00
Statutorily protected species	40	7.37	8.00	25	6.96	8.00	15	8.07	9.00
Rare or threatened ecological, hydrological or geomorphological processes	40	7.95	8.00	25	7.80	8.00	15	8.20	10.00
Diversity/richness									
Freshwater plants	43	7.14	7.00	27	6.93	7.00	16	7.50	8.00
Freshwater invertebrates	43	7.28*	8.00	27	6.81	7.00	16	8.06*	8.50
Freshwater fish	43	7.16	7.00	27	6.70	7.00	16	7.94	8.50
In-lake habitats	43	7.21*	8.00	27	6.78	7.00	16	7.94*	8.50
Adjacent habitats	43	6.72**	7.00	27	6.15	7.00	16	7.69**	8.00
Special features									
Important floodplains and/or adjacent wetland habitats (e.g. fen, marsh, wet woodland)	42	7.55**	8.00	27	6.89	8.00	15	8.73**	10.00
Presence of economically important species	42	4.95	5.00	27	4.37	4.00	15	6.00	7.00

* $p < 0.05$
** $p < 0.01$
*** $p < 0.001$

Table 2.6 *Mean and median criterion scores for rivers, analysed in three broad groups: A: Universities, research institutes, B: Conservation bodies, C: River, lake and land managers/ environmental regulators*

	A			B			C		
Criteria	*N*	Mean	Median	*N*	Mean	Median	*N*	Mean	Median
Naturalness	23	7.92	7.75	12	8.19	8.25	26	7.99	8.00
Rarity	23	8.12	8.43	12	7.04	6.86	26	8.01	8.14
Representativeness	23	7.05	7.50	12	6.15	6.67	26	6.87	7.25
Diversity/richness	23	6.98	7.20	12	5.83	5.50	26	7.42	8.00
Special features	23	6.91	7.00	12	4.67	5.25	26	6.58	6.25

Table 2.7 *Mean and median criterion scores for lakes, analysed in three broad groups: A: Universities, research institutes, B: Conservation bodies, C: River, lake and land managers/ environmental regulators*

	A			B			C		
Criteria	*N*	Mean	Median	*N*	Mean	Median	*N*	Mean	Median
Naturalness	11	7.94	7.88	7	8.20	8.00	15	7.91	8.00
Representativeness	11	6.91	6.83	7	5.21	5.33	15	7.10	7.17
Rarity	11	8.13	8.29	7	7.27	7.00	15	8.05	8.14
Diversity/richness	11	7.18	7.00	7	6.06	5.40	15	7.83	8.00
Special features	11	6.59	7.00	7	3.64	4.00	15	6.90	8.00

Despite general similarities between scores given by UK and US respondents, the pattern of higher mean scores assigned by US respondents was consistent, striking, and strangely puzzling. There seems no reason why freshwater specialists in the USA should consistently score conservation attributes more highly than their UK counterparts; perhaps the observed pattern reflects some fundamental cultural and behavioural differences rather than anything directly related to fresh waters and their conservation.

The 11 attributes where statistically significant differences were found among the three professional groupings showed one particularly interesting feature: 7 of them were within the categories of rarity or diversity/richness. Perhaps one explanation may be a greater awareness by those whose full-time work is in conservation; in other words, the characteristics of rarity and species diversity (so often promoted to the

Table 2.8 *Eleven attributes (summary descriptions) for which there were significant differences in scores allocated by groups A, B and C*

Criterion	Attribute	Significance (*p*)
Rivers – Rarity	Statutorily protected species	0.014
Rivers – Diversity/richness	Freshwater fish	0.006
Rivers – Diversity/richness	Instream habitats	0.013
Rivers – Diversity/richness	Adjacent habitats	0.001
Rivers – Special features	Floodplain/wetland habitats	0.006
Lakes – Representativeness	Physical features	0.049
Lakes – Rarity	Endemic species	0.029
Lakes – Diversity/richness	Freshwater fish	0.009
Lakes – Diversity/richness	Adjacent habitats	0.013
Lakes – Special features	Floodplain/wetland habitats	0.019
Lakes – Special features	Economically important species	0.029

Table 2.9 *Overall criterion indices for all categories combined, compared with a similar study in Australia (Dunn, 2004)*

Criteria	USA/UK	Australia
Naturalness	8.0	7.9
Rarity	7.9	7.8
Diversity/richness	7.0	7.0
Representativeness	6.7	6.8

public as the chief concerns of conservation) are merely one element of a much larger canvas.

This study was designed partly to match the work carried out recently in Australia on assessing the conservation value of rivers (Dunn, 2004). Although the raw data of the Australian study are not available for statistical comparison, there is a remarkable similarity between the mean criterion scores in the US/UK study and those reported from Australia (Table 2.9). The overall order of importance of the four criteria is the same, with the mean scores almost identical. A more detailed comparison of the two studies is given in Table 2.10, showing once again the general similarity in the results of the two studies.

The results of the exercise suggest that perceptions of conservation value can be readily summarized using the traditionally accepted criteria of naturalness, rarity, diversity and representativeness. The process by

Table 2.10 *A comparison of conclusions stated by Dunn (2004) compared with the present study (river responses only)*

Dunn(2004)	UK/US study
Some used full range (0/1–10), others only more than 6 or 7	YES. Out of 72, max 10 = 58; min 0 = 9; full 0–10 = 7; min ≥6 = 13 No. of respondents using 1–11 points on 11-point scale: 1 = 0; 2 = 0; 3 = 2; 4 = 9; 5 = 4; 6 = 16; 7 = 15; 8 = 9; 9 = 7; 10 = 3; 11 = 7
Every attribute achieved at least some scores of 10	YES
11 attributes had at least one rating of 0	8 attributes with at least one 0
No attribute had a notably low overall rating	YES
Lower mean ratings tended to have higher variance than attributes with mean ratings of 8 or above	YES. Negative correlation between high variance and low mean
Ratings of most attributes varied over the whole range from 0/1–10 – only exceptions: intact riparian vegetation (4–10), rare or threatened communities or ecosystems (5–10), rare or threatened habitats (4–10)	NO. Of 28 named attributes, only 13 covered the range 0/1–10
Four individuals rated all attributes under Representativeness as uncertain	Two individuals rated all attributes under Representativeness as uncertain
Four individuals rated all attributes under Naturalness as 10, whereas most used a range of scores	Three individuals rated all attributes under Naturalness as 10, whereas most used a range of scores
Overall mean scores suggest general support rather than support from small number of specialists	YES
Even attributes with highest scores overall were considered to be of minor importance by a few individuals	YES
Seven of the top 10 scores were for attributes in the naturalness criterion	Seven of the top 11 scores (two exactly equal) were for attributes in the naturalness criterion
Lowest mean scoring attributes attracted more 'uncertain' responses and displayed more variance	YES
Three of the four representativeness attributes fell amongst the lowest scores	YES. All six of the representativeness attributes fell amongst the 10 lowest scores
Even the lowest scoring attributes were considered very important by a few individuals who gave them high scores	YES
Overall, naturalness was considered to be the most important criterion (7.9), closely followed by rarity (7.8)	Overall, naturalness was considered to be the most important criterion (8.0), closely followed by rarity (7.9)

which evaluations are made may not always fall neatly into these categories. Yet it is apparent from the ensuing chapters in this book that when rivers and lakes are selected for special protection, when legal challenges are made to development projects likely to damage freshwater ecosystems, or when plans are made to restore degraded water bodies, it is some combination of these criteria that will often be applied to assess conservation value.

Acknowledgements

We thank Jonathan Higgins, Catherine Duigan, Mary Freeman, Margaret Palmer and Laurie Duker for their contributions.

References

Allen, D. E. (1976). *The Naturalist in Britain: A Social History*. London: Allen Lane.

Benson, M. G. (1970). *A Century of Fisheries in North America*. Special Publications 7. Bethesda, MD: American Fisheries Society.

Boon, P. J. (1994). Nature conservation. In *The Fresh Waters of Scotland: A National Resource of International Significance*, eds. P. S. Maitland, P. J. Boon & D. S. McLusky. Chichester: John Wiley, pp. 555–76.

Boon, P. J. (2000). The development of integrated methods for assessing river conservation value. *Hydrobiologia*, **422/423**, 413–28.

Boon, P. J. & Howell, D. L. (eds.) (1997). *Freshwater Quality: Defining the Indefinable?* Edinburgh: The Stationery Office.

Boon, P. J. & Lee, A. S. L. (2005). Falling through the cracks: are European directives and international conventions the panacea for freshwater nature conservation? *Freshwater Forum*, **24**, 24–37.

Boon, P. J., Holmes, N. T. H., Maitland, P. S. & Fozzard, I. R. (2002). Developing a new version of SERCON (System for Evaluating Rivers for Conservation). *Aquatic Conservation: Marine and Freshwater Ecosystems*, **12**, 439–55.

Boon, P. J., Holmes, N. T. H., Maitland, P. S., Rowell, T. A. & Davies, J. (1997). A system for evaluating rivers for conservation ('SERCON'): development, structure and function. In *Freshwater Quality: Defining the Indefinable?*, eds. P. J. Boon & D. L. Howell. Edinburgh: The Stationery Office, pp. 299–326.

Brouha, P. (1993). The emerging science-based advocacy role of the American Fisheries Society. *Journal of the North American Benthological Society*, **12**, 215–16.

Calicott, J. B. (1990). Whither conservation ethics? *Conservation Biology*, **4**, 15–20.

Christoffersen, L. E. (1997). IUCN: a bridge-builder for nature conservation. In *Green Globe Yearbook 1997*, eds. H. O. Bergesen & G. Parmann. Lysaker: The Fridtjof Nansen Institute, pp. 59–69.

Dewberry, T. C. & Pringle, C. M. (1994). Lotic science and conservation: moving toward common ground. *Journal of the North American Benthological Society*, **13**, 399–404.

Dunn, H. (2004). Defining the ecological values of rivers: the views of Australian river scientists and managers. *Aquatic Conservation: Marine and Freshwater Ecosystems*, **14**, 413–33.

Evans, D. (1992). *A History of Nature Conservation in Britain*. London: Chichester.

Forestry Commission (2003). *Forests & Water Guidelines*. 4th edn. Edinburgh: Forestry Commission.

Freshwater Biological Association (FBA) (2001a). *Lake Assessment and the EC Water Framework Directive*. Special issue of *Freshwater Forum*, Vol. 16. Ambleside: Freshwater Biological Association.

Freshwater Biological Association (FBA) (2001b). *European Temporary Ponds*. Special issue of *Freshwater Forum*, Vol. 17. Ambleside: Freshwater Biological Association.

Howell, D. L. (1994). Role of environmental agencies. In *The Fresh Waters of Scotland: A National Resource of International Significance*, eds. P. S. Maitland, P. J. Boon & D. S. McLusky. Chichester: John Wiley, pp. 577–611.

Macan, T. T. (1951). *Life in Lakes and Rivers*. The New Naturalist. London: Collins.

Naiman, R. J., Magnuson, J. J., McKnight, D. M. & Stanford, J. A. (1995). *The Freshwater Imperative: A Research Agenda*. Washington, DC: Island Press.

National Research Council (NRC) (1992). *Restoration of Aquatic Ecosysems: Science, Technology, and Public Policy*. Washington, DC: National Academies Press.

National Research Council (NRC) (2000). *Watershed Management for Potable Water Supply: Assessing the New York City Strategy*. Washington, DC: National Academies Press.

National Research Council (NRC) (2001a). *Envisioning the Agenda for Water Resources Research in the Twenty-first Century*. Washington, DC: National Academies Press.

National Research Council (NRC) (2001b). *Assessing the TMDL Approach to Water Quality Management*. Washington, DC: National Academies Press.

National Research Council (NRC) (2002a). *Riparian Areas: Functions and Strategies for Management*. Washington, DC: National Academies Press.

National Research Council (NRC) (2002b). *The Missouri River Ecosystem: Exploring Prospect for Recovery*. Washington, DC: National Academies Press.

National Research Council (NRC) (2004). *Managing the Columbia River: Instream Flows, Water Withdrawals and Salmon Survival*. Washington, DC: National Academies Press.

National Research Council (NRC) (2005). *Valuing Ecosystem Services: Towards Better Environmental Decision-making*. Washington, DC: National Academies Press.

Nature Conservancy (1972). Management of wildlife as a natural resource. Views of the Nature Conservancy. Paper prepared by the UK Working Party on the Management of Natural Resources, for the UN Conference on the Human Environment, Stockholm, June 1972. Huntingdon: Nature Conservancy.

Nature Conservancy Council (NCC) (1975). *First Annual Report*. London: NCC.

Nature Conservancy Council (NCC) (1984). *Nature Conservation in Great Britain*. Peterborough: NCC.

Nature Conservancy Council (NCC) (1989). *Guidelines for Selection of Biological SSSIs*. Peterborough: NCC.

Nicholson, M. (1970). *The Environmental Revolution*. London: Hodder and Stoughton.

Nicholson, M. (1987). *The New Environmental Age*. Cambridge: Cambridge University Press.

Palmer, T. (1994). *Lifelines: The Case for River Conservation*. Washington, DC: Island Press.

Pringle, C. M. & Aumen, N. G. (1993). Current issues in freshwater conservation: introduction to a symposium. *Journal of the North American Benthological Society*, **12**, 174–6.

Pringle, C. M. Rabeni, C. F., Benke, A. & Aumen, N. G. (1993). The role of aquatic science in freshwater conservation: cooperation between the North American Benthological Society and organizations for conservation and resource management. *Journal of the North American Benthological Society*, **12**, 177–84.

Ratcliffe, D. A. (1977). Nature conservation: aims, methods and achievements. *Proceedings of the Royal Society of London, Series B*, **197**, 11–29.

Rowan, J. S., Carwardine, J., Duck, R. W. *et al.* (2006). Development of a technique for Lake Habitat Survey (LHS) with applications for the Water Framework Directive. *Aquatic Conservation: Marine and Freshwater Ecosystems*, **16**, 637–57.

Sheail, J. (1998). *Nature Conservation in Britain: The Formative Years*. London: The Stationery Office.

3 · *Freshwater conservation in action: contrasting approaches in the USA and the UK*

CATHERINE M. PRINGLE AND DAVID WITHRINGTON

Introduction

The protection of freshwater ecosystems is a formidable challenge in the human-dominated landscapes of the USA and the UK. Both countries are grappling with how to balance trade-offs between human use of water resources and protection of the natural functioning of freshwater ecosystems. Due to changes in public values and legislation, including the perceived effects of climate change, the allocation of water to maintain the ecological value of biota and freshwater ecosystems is beginning to be considered in addition to water allocation for agriculture, urban and industrial use. Differences in conservation approaches in the two countries are a reflection of differences in the history of human settlement, various socio-political factors, geographic extent (size), climate and land use.

The UK has had a much longer period of intensive human settlement than the USA. By the twelfth century its native forests had been substantially destroyed. The Magna Carta (1215) gave the government control of managing wildlife for the public good in England, while the English king and Parliament retained hunting privileges. One result was the removal of mill weirs from a few rivers to allow for fish passage (although far more mill weirs were created than removed). Regulatory directives for water pollution control stem from the fourteenth century. The earliest recorded law seeking to regulate this problem in England was the Act for Punishing Nuisances which Cause Corruption of Air near Cities and Great Towns of 1388. Its preamble speaks of 'so much dung and other filth of the garbage and entrails as well as of beasts killed, as of other corruptions, be cast and put in ditches, rivers and other waters'. An Act for the Preservation of the River Thames in 1535 made it an offence to cast dung, rubbish or other things in that river.

While the tradition of governmental control over fish and wildlife was taken to the USA, the destruction of habitat was widespread during

Assessing the Conservation Value of Fresh Waters, ed. Philip J. Boon and Catherine M. Pringle. Published by Cambridge University Press.

colonization, particularly following the draining of wetlands authorized under the Louisiana Swamp Land Act of 1849 (and Swamp Land Acts of 1850 and 1860). Forests in the USA were extensively exploited for lumber and agriculture during the colonial period in the late sixteenth and early seventeenth centuries. While all US states had adopted laws that regulated hunting and fishing by 1890, these efforts unfortunately often had little influence on the damaging environmental effects of water development projects (e.g. flood control, navigation and irrigation).

In both the USA and the UK, early steps in controlling water pollution were often motivated by concern about waterborne diseases (such as infectious hepatitis, cholera, bacterial dysentery and giardiasis) that were common before the twentieth century. In the UK, much of the history of the Middle Ages revolves around disease and plague. Similarly, cities in the eastern USA suffered from cholera and typhoid epidemics due to contaminated water supplies in the late seventeenth and early eighteenth centuries. The now famous study which made the scientific connection between drinking water, water pollution and diseases was made in England by Dr John Snow in 1854, when he found that cholera cases in London were clustered in an area served by a community water pump (Snow, 1936). It is interesting to note that historic concern over drinking water quality in both the USA and the UK did not lead initially to measures involving the clean-up or protection of freshwater ecosystems. Instead, deaths from waterborne disease were avoided by abandoning contaminated drinking water sources and using alternative ones, as well as by adopting chlorination and water filtration.

Today, with human populations in the UK and the USA at 60 and 298 million respectively, these approaches are no longer sustainable. Various legislative tools and state and central government agencies have been created for the management and protection of water resources. The balance between economic development and environmental protection of freshwater ecosystems – for sustainable human use and for natural values – has become a major challenge, and a more holistic ecosystem approach is being embraced by politicians, although the practical implementation of appropriate measures has been slow to emerge.

This chapter explores freshwater conservation in action in the USA and the UK, building on Chapter 2 which provides the philosophical context for freshwater conservation and background information on water resource legislation and agency structure in both countries. We discuss current conservation approaches and activities in the USA and the UK, and provide examples which serve to illustrate similarities and differences in conservation challenges and approaches taken in the two countries.

Contrasts in the regulatory influence of international versus national legislation

While international legislation has played a central role in the UK, in terms of protecting the environment and developing freshwater conservation for natural values, the USA has been considerably less influenced by international agreements.

For political and economic reasons, the USA has been far less participatory than the UK in international agreements involving protection of the environment. For example, the USA has not signed the Convention on Biodiversity (CBD), which entered into force in 1993 and was negotiated under the auspices of the United Nations Environment Programme. By 1998, more than 170 countries had become signatories. The three goals of the CBD are to promote the conservation of biodiversity, the sustainable use of its components, and the fair and equitable sharing of benefits arising out of the utilization of genetic resources. Likewise, the Kyoto Protocol, negotiated by more than 160 nations in 1997, aims to reduce net emissions of certain greenhouse gases (primarily CO_2). While the Protocol has been celebrated as a historic breakthrough by governments of Western Europe, the USA is yet to sign it.

Under the 1971 Ramsar Convention, signatories designate wetland sites of international importance especially for the protection of waterfowl habitat. Estimates of the total area of wetlands that have been designated in both the USA and the UK are 1 300 000 and 920 000 ha, respectively. These figures are surprisingly close, given that the USA is nearly 40 times larger than the UK. Similarly, it is interesting to note that 168 Ramsar sites have been designated in the UK, while only 24 sites have been designated in the USA, again surprising, given the disparity in size between the two countries.

European Union environmental legislation

International legislation emanating from the European Union (EU) has had a major influence on freshwater conservation in the UK and other Member States over the past 20 years. The European Commission (EC), which has its headquarters in Brussels, consists of commissioners from the

Member States, supported by civil servants. Its role is to draft proposals to implement Union policy, to ensure that the EU legislation is applied correctly by Member States and to administer the budget. The limits of the Commission's authority are clearly defined, and legislative decisions in the form of directives are taken by the Council of Ministers of the 27 Member States and the elected European Parliament. Key legislation from the EU driving freshwater conservation in the UK includes the Freshwater Fish Directive (1978), the Urban Wastewater Treatment Directive (1991), the Habitats Directive (1992) and the Water Framework Directive (WFD; 2000). The Urban Wastewater Treatment Directive and the WFD are discussed below, followed by an account of how the EC enforces legislation. (See Chapter 2 for more detailed descriptions of EU legislative acts.)

The Urban Wastewater Treatment Directive has been pivotal in shaping current approaches to wastewater treatment in the UK. Until the 1980s, phosphorus was not removed from any sewage effluent, and was causing widespread eutrophication of rivers, lakes and shallow coastal waters in the UK. Tertiary treatment, using ferric salts to precipitate out phosphate, was installed at a few sewage treatment works (STWs) discharging into rivers in Norfolk – and contributing to enrichment of the shallow lake system of famous wildlife refuges in the Norfolk Broads – and at Hawkshead, a village at the head of Esthwaite Water in the Cumbrian Lake District.

The Urban Waste Water Treatment Directive required the designation of Sensitive Areas at risk from eutrophication and removal of phosphorus at contributing STWs by 1998. In 1989, water authorities were privatized in England and Wales and became regional water companies, charging water consumers for their services – water supply and sewage disposal. This gave the government an opportunity to set price limits and five-yearly investment programmes. In the period 1995–2000 – known as Asset Management Programme 2 (AMP2) – removal of phosphorus or nitrogen was implemented at STWs affecting 77 Sensitive Areas to meet the requirements of the EC Directive. A major breakthrough came in 2000, when English Nature (now Natural England), assisted by political lobbying from wildlife NGOs (non-governmental organizations), secured a programme under AMP3 to remove phosphorus at 65 STWs to protect 32 Sites of Special Scientific Interest (SSSIs) in England from eutrophication at a cost of £100 million. In AMP4, English Nature obtained a further programme of water quality improvements in waters designated for nature conservation, mainly under the EU Habitats Directive, in the period 2005–2010, amounting to a capital investment of £448 million. This is the largest financial investment in the history of nature conservation in the UK.

The aim of the WFD is that all surface waters achieve 'good ecological status' by 2015. Further information on the WFD is given in Chapter 2. Although implementation of the WFD potentially provides a sustainable future for the UK's lakes and rivers, much depends on the standards that are agreed for good ecological status and on the exemptions that are invoked on the grounds of cost. The risk assessment carried out for the WFD in 2004 has revealed the potential costs of dealing with the legacy of centuries of environmental degradation in the UK (www.defra.gov.uk/environment/water/wfd/riverbasincharacterisation).

Another important aspect of WFD implementation is the coverage of the Directive and its River Basin Management Plans (RBMPs). Answers by UK government ministers to questions in Parliament on 18 January 2006 revealed that, in the initial river basin characterization for England and Wales in 2005, less than a third of the river network and only 7% of lakes were identified by the Environment Agency as 'water bodies' eligible for WFD measures. In 2005, the UK government announced a further phase of water body identification up to 2007 (Defra *et al.*, 2005). In England and Wales, the conservation agencies have proposed lists of additional surface waters of significance for biodiversity and recommended that RBMPs should cover the whole river network.

The European Commission has the power to act against a Member State in enforcing EU law, for example by withholding payment of European development funds, if the Member State does not implement a judgement made by the European Court of Justice. At the end of 2004, 2531 cases against Member States were outstanding, about a third of which related to 'environmental' directives (www.europa.eu.int/comm/secretariatgeneral/sgb/droit_com). At the end of 2005, procedures were open against Belgium, Denmark, France, Germany, Greece, the Netherlands, Luxembourg and Spain for failure to apply the 1996 Directive on Integrated Pollution Prevention and Control limiting harmful emissions to air, water and land from major industrial operations. The Commission has sent final warnings to Italy, Spain and Greece for not complying with basic provisions under the WFD. The UK was referred by the Commission to the European Court of Justice on 13 December 2005 under the 1991 Waste Water Treatment Directive for failing to install secondary treatment at 14 coastal sewage discharges. The UK has a reputation for diligent application of EU directives, and there is an active NGO community in the UK to provide information on cases to the Commission however, the government appears to be in denial about certain aspects of pollution in coastal waters.

Conflicts between EU environmental legislation and regulatory frameworks in the USA

The potential for conflict can occur when international legislation from the EU differs from the regulatory framework used in other parts of the world. An ongoing example concerns the EU legislation called REACH (Registration, Evaluation and Authorization of Chemicals) designed to protect human health and the environment from harmful chemicals. This legislation requires chemical producers and importers to register chemicals in a central database along with information regarding their safe use. The aim is to improve protection through better and earlier identification of the properties of chemical substances. The REACH proposal gives greater responsibility to industry to manage risks from chemicals and to provide safety information on the substances.

This progressive and environmentally important legislation has been subject to some of the most intensive lobbying opposition (by the USA) that the EU has ever experienced. While it has previously been US policy not to interfere with another country's efforts to protect its own environment, a disturbing reality is that the US chemical industry successfully lobbied the US government to block REACH, which has resulted in the weakening of this important environmental initiative (Waxman Report, 2004). Arguments made by the chemical industry against REACH were that it would interfere with trade, increase costs, discourage innovation and hamper commerce. In opposing REACH, the chemical industry recommends voluntary measures and using data and information already available instead of requiring testing. The REACH initiative contrasts markedly with the main legislative tool used in the USA to regulate chemicals, the 1976 Toxic Substances Control Act, which is thought by many to be weak and too deferential to industry.

REACH legislation, as originally enacted, would fundamentally change safety laws, forcing manufacturers to conduct extensive tests on chemicals they want to use or make. Proposed regulatory changes support the 'precautionary principle', a philosophy that says that if science is uncertain of whether a substance is harmful, it should not be allowed. Before international pressure weakened REACH, it would have required that chemicals found to be dangerous would be phased out within 5 years unless their continued use had some strong social or economic justification.

Historical changes in the mission of government agencies

In both the USA and the UK, there has been a changing emphasis by government agencies from management of freshwater resources for utilitarian values towards protection and restoration for natural values, which is discussed below.

USA

The water resources infrastructure in the USA (Figure 2.1) was clearly not designed to consider ecosystem water needs. Instead, it was created to minimize damage from flooding, and to provide water for municipalities, navigation and the generation of energy. Water resource agencies of the US federal government were created to promote the settlement of frontier territories. After the Second World War, the federal focus shifted from water development to environmental protection and conservation. New agencies were created such as the US Environmental Protection Agency (EPA), while the missions of other agencies (e.g. the US Fish and Wildlife Service (FWS)) were expanded. Some of these new missions contrasted greatly with the historic mission of development agencies and the consequence has been conflict, sometimes between federal agencies owing to opposing federal missions, and at other times between existing federal policy and diverging constituencies. Consequently, the implementation of freshwater conservation objectives in the USA depends on complex (and sometimes contentious) interactions between federal and individual state government agencies.

The US Army Corps of Engineers (USACE) provides a good example of historical changes in the mandates of a federal water resource agency, illustrating expanding governmental emphasis on freshwater conservation for natural values. The Corps is the oldest water resource management agency in the USA, which has historically dealt with flood control and navigation improvement. Clean Water Act amendments of 1972 gave the US Army Corps and the EPA responsibility for protecting wetlands in the USA, and environmental restoration is a more recent addition to its mission.

Another example is the US Bureau of Reclamation, which has been the premier water development agency in the USA since the early 1900s – in charge of the development of irrigation projects to promote settlement of the arid western USA. In 1993, the Bureau declared that now that it had achieved its goals of increasing settlement in the West its new mission would be to manage existing water projects to promote conservation and to develop partnerships with customers, states and native American tribes. The Bureau of Reclamation has been strongly affected by the anti-dam construction movement in the USA and has transferred ownership and operation of many dams and water delivery systems to local water agencies.

The US Bureau of Reclamation's shift in policy towards environmental protection is also illustrated by its contentious decision in 2001 not to deliver water to an irrigated area in the western USA. The decision was made in order to protect instream flows in the Klamath River. While it protected the biological integrity of the Klamath River, it left farmers without water, and their crops died. The Bureau decided that because of the requirements of the Endangered Species Act it would not be able to deliver water from Upper Klamath Lake to irrigators receiving water from the Klamath Project during a 'critically dry year'. The Klamath Project was authorized by Congress in 1905 to deliver irrigation water to the dry lands of Southern Oregon, an area that receives only 15–20 cm of average annual precipitation. The Project serves ~1400 farms and provides irrigation water to ~85 000 ha of land which is farmed for alfalfa, barley, oats, wheat, potato and sugar beet (Cech, 2005). Retention of water in Upper Klamath Lake provided environmental benefits to the Lost River sucker (*Deltistes luxatus*), the shortnose sucker (*Chamistes brevirostris*) and the Southern Oregon/Northern California coasts' coho salmon (*Oncorhynchus kisutch*).

UK

Over the last 50 years, there have been constant changes in the structure and missions of government bodies dealing with nature conservation and the management of the water environment. The conservation agencies began in 1949 as the Nature Conservancy (NC) covering the whole of Great Britain. Northern Ireland always had separate arrangements. In 1974, the NC was split from its research arm and renamed the Nature Conservancy Council (NCC). In 1991, three new bodies – Scottish Natural Heritage (SNH), Countryside Council for Wales (CCW) and English Nature – were created, following concerns that the NCC had been too vociferous in opposing commercial afforestation with conifers of the greatest remaining peat bogs in the Flowe country of north-east Scotland. Ironically, the government's Forestry Commission announced a new policy direction in 2005, whereby 'millions of conifers and non-native trees will be removed' over the next 20 years, at least in England.

Both SNH and CCW have responsibility for landscape and outdoor recreation, in addition to nature conservation. In October 2006, English Nature was merged – with the Countryside Agency's landscape and recreation functions and with the government's Rural Development Service, which negotiates and processes environmental agreements with farmers – to become Natural England. At the same time, CCW is relinquishing its role in agri-environment agreements back to central government in the Welsh Assembly. There have been several changes in direction and organizational structure, and no doubt there will be more to come, as successive governments wish to be seen to be doing something different in the environmental field.

Changes in the missions of agencies responsible for managing and regulating water resources in the UK have been frequent and subject to structural reorganization. For example, the Environment Agency for England and Wales has its roots in the regional water authorities of the 1960s and the river conservancies, which were responsible for land drainage. At that time there was no central agency, and these authorities were managed directly by government departments. In 1989, the Conservative government decided to sell off the water authorities – responsible for public water supply and sewage disposal – to the highest bidders, but they were given duties to further the conservation of special sites. Almost as an afterthought, a National Rivers Authority was set up in 1990 to regulate abstraction, discharges and fisheries and to undertake flood management. These activities had different Acts of Parliament and regulatory regimes, responding to and funded by different government departments: the Department of the Environment had policy responsibility for water resources and abstraction, while the Ministry of Agriculture funded fisheries and flood defence.

In 1996, the National Rivers Authority was merged into a larger Environment Agency, incorporating Her Majesty's Inspectorate of Pollution from central government, dealing with air and major industrial emissions, and the waste regulation authorities from local government. The present-day Environment Agency is the competent authority for implementing the WFD in England and Wales; however, its responsibilities span a much wider field of activity, both as regulator and as provider of flood defences. In the meantime, flood management, fisheries, water resources and water quality have been housed together in one government department – Defra (Department for Environment, Food and Rural Affairs).

With so many different and sometimes conflicting missions, it has been difficult for the Environment Agency to deliver programmes, especially in the area of nature conservation, which is not one of its core missions. Is the Agency a regulator or a facilitator, or is it seeking a balance between its various functions? Has it become too internally focused? One thing that seems fairly certain is that, with the historical frequency of structural changes, more are in the offing.

The duality of mandates within and between government organizations

In the USA, and to a lesser extent in the UK, government agencies involved in freshwater resource issues are challenged by a duality of mandates. To a degree this is the result of giving them additional functions, rather than creating new single-issue agencies.

USA

This duality of mandates provides a difficult balancing act within and between government agencies (Cech, 2005), which are often charged with reconciling conflicts between economic development and environmental protection. Environmental efforts within agencies can be hindered by the practical implementation of concepts such as 'sustainability' and 'ecosystem management' which are still only vaguely defined. From the perspective of a resource manager, it is necessary to define (with scientific support) what the limits are on how much human alteration a freshwater ecosystem can sustain, and in most cases such limits have not been defined. US government agencies also face constraints of limited funding and personnel, and the politicization of natural resources decision-making is causing increasing conflicts at all levels of government (Cech, 2005).

Dozens of different government entities are involved in deciding which wastes can be discharged into water and how water is used and redistributed. Often, the goals of one agency are at cross-purposes with those of others (Baron *et al.*, 2003). Moreover, US laws and regulations concerning water resources are often implemented in a management context that focuses primarily on maintaining the lowest acceptable water quality and minimal flows for natural ecosystem function (Baron *et al.*, 2003).

In the USA, different categories of public lands are managed for different types of multiple uses by specific federal agencies. For example, the US National Park Service manages national parks; US Department of Agriculture Forest Service manages national forests; and US FWS manages national wildlife refuges. The missions of these agencies involve managing these lands for multiple uses, seeking to balance environmental with public needs. The US Forest Service is the federal agency charged with managing national forests and associated water resources. National Forests in the USA provide $>2.4 \times 10^6$ ha of wetlands for waterfowl and wetland-associated wildlife. Waters draining national forests fulfil multiple purposes in addition to wildlife and other environmental needs, including human water supplies, domestic livestock and timber management. In fact, as much as half of the human population in the eastern USA derive water supplies from national forest catchments.

A tradition and policy of multiple resource use and development on national forest land has allowed many private and non-federal water users to build water facilities and withdraw water from within national forests. It is now clear that many water development facilities located within national forests are linked to deteriorating aquatic resources such as fish populations. In some cases, the Forest Service is now issuing permits that require owners of dams and other facilities to leave enough water in national forest streams to sustain fish during times of low water flows. Controversies arise when facility owners object to these conditions and have led the Forest Service to develop a special Water Rights Task Force to examine conflicts between the US Forest Service and the water users.

UK

There is little overlap between agencies for controlling water pollution and abstraction in the UK, with one authority (Environment Agency in England and Wales and Scottish Environment Protection Agency (SEPA) in Scotland) being responsible. However, when activities affect SSSIs, the conservation agencies also have a consenting function.

While the Environment Agency might be expected to uphold the protection of SSSIs, there can be conflicts (e.g. between its role as a land drainage authority and its duty towards nature conservation). One reason given by the Agency for granting landowners permission in 2006 to remove gravel from the bed of the River Eden SSSI (which is also designated as a European Special Area of Conservation) was because its land drainage by-laws – which date back to the 1970s – do not take account of conservation, but only of flooding. Another area where the Environment Agency has experienced an apparent conflict of interests is between its duty to ensure that water companies can provide an adequate public water supply and its duty to further the conservation of SSSIs. Abstraction licences are financially valuable, and in the drier parts of the

UK water resources are almost fully utilized. It is in these areas where wetlands are affected most by abstraction, usually from groundwater. The Environment Agency's reluctance over the years to amend licences for public supply was further compounded by the removal of abstraction from the five-yearly investment programme of water companies in England and Wales for 2005–2010. Any amendments to licences would be open to appeal and payment of compensation to water companies by the Environment Agency was estimated at over £84 million on priority EU Habitats Directive sites alone. For 2010–2015, abstraction affecting Habitats Directive sites has been reinstated in the water companies programme, but all other licence changes would remain subject to compensation. In 2008, the government published proposals (Defra, 2008) to impose a time limit on all abstraction licences, after which they would be subject to renegotiation without compensation.

Freshwater conservation is not always the lowest priority, and there are more and more situations where there is a positive outcome. This appears to be because the laws are strict – for instance, the test in the Habitats Directive is to show that there is 'no adverse effect on the integrity' of a nature conservation site, the burden of proof being on the developer. National laws, such as the Wildlife and Countryside Act, are reflected in the government's public service agreement for SSSIs, where a favourable condition as defined by Natural England is to be reached by 2010 for 95% of sites in England (a target date of 2015 was set for Wales in 2006). In England, river and stream SSSIs are the habitat category in worst condition with 29% in target condition (English Nature, 2006). So there is a real challenge ahead.

Cooperative conservation: partnerships between government and non-government agencies and stakeholders

Increasingly, government organizations in both the USA and the UK have taken the approach of forging partnerships with stakeholders (representing many segments of society) to address freshwater resource issues in a given region more effectively.

USA

In the USA, cooperative conservation involving partnerships has become the cornerstone for successful conservation efforts. For example, the USACE has recently partnered with the Nature Conservancy (TNC, a non-government agency) to improve the management of dams on various rivers across the country. Under the new partnership, the Sustainable Rivers Project, the two organizations are working together to help restore and protect the health of rivers and surrounding natural areas, while continuing to meet human needs for flood control and power generation services. One focus of efforts will be in the Green River where 35 aquatic species are considered to be imperilled. The TNC is working with the USACE to alter their operation of the Green River Dam (located at the river's headwaters) to restore natural flow regimes to benefit wildlife (nature.org/initiatives/freshwater/work/greenriver.html).

Another example involves partnerships forged between multiple government agencies and stakeholders to protect the ivory-billed woodpecker (*Campephilus principalis*) in the south-eastern US state of Arkansas. The US FWS is charged with the protection of endangered species and associated habitat. The FWS has developed a Cooperative Conservation Initiative, which includes a variety of grant and technical aid programmes to support wildlife recovery. In response to the dramatic rediscovery of the ivory-billed woodpecker in 2004 at the Cache River National Wildlife Refuge in Arkansas, the FWS and local citizens are working to develop a 'Corridor of Hope Cooperative Conservation Plan' to save the woodpecker. This Corridor of Hope refers to a 120 mile long × 20 mile wide swath of land (much of it wetland) in Arkansas where the woodpecker has been sighted. Plans will be developed and implemented whereby local citizens participate in writing a recovery plan that maintains historic public uses of land while protecting the bird's habitat.

A multimillion-dollar partnership was formed to support this effort: the Interior Department, along with the Department of Agriculture, has proposed that more than $10 million in federal funds be committed to protect the bird. This amount supplements the $10 million already committed to research and habitat protection efforts by private sector groups and citizens. Federal funds will be used for research and monitoring, recovery planning and public education. In addition, the funds will be used to enhance law enforcement and conserve habitat through conservation easements, safe-harbour agreements and conservation reserves. Similar innovation and partnership efforts have proved very effective in the recovery of whooping cranes (*Grus americana*) that were nearly extinct.

UK

In the UK, the majority of recent conservation projects are partnerships. The biggest grant-giving body, the Heritage Lottery Fund, welcomes applications from organizations working together and manages a special Landscape Partnerships Fund. Conservation challenges are often complex and on landscape scales, involving conserving water resources and endangered species across an entire region, requiring the participation of local communities, industry, private groups and other government agencies. Partnerships between the public, private and government sectors to achieve freshwater conservation are also increasing in the UK.

As one example, a '50-year vision for England's water and wetlands' was launched at World Wetlands Day in London in February 2006. This is a partnership venture between the two major voluntary bodies – the Royal Society for Protection of Birds (RSPB) and the Wildlife Trusts – and the government agencies – Natural England, English Heritage and the Environment Agency. The 'vision' outlines the scale of wetland creation and restoration that is needed to meet conservation objectives. It aims to identify the most suitable locations for different types of wetland, through the development of a Geographic Information System tool, which will also be applicable at local level.

In 2004, a pilot project was set up to test how changes to land-use management on a catchment scale could contribute to reducing flood risk and at the same time bring biodiversity benefits. The project was based in the catchment of two small rivers – the Skell and the Laver – upstream of the Yorkshire city of Ripon, which experienced flooding in 2000. It was a partnership between Defra, the Environment Agency, Natural England, the government's Rural Development Service, the Forestry Commission and two private bodies – RSPB and the Country Land and Business Association. Because of cuts in government funding, the project was closed in 2006, but the Forestry Commission is continuing to research the potential benefit of woodland planting and management in the floodplain on flood attenuation, and three universities are carrying out related research projects in the catchment. The partnership is continuing in order to evaluate what has been achieved and, if possible, to commission further work. One lesson learnt is that, while the partners will contribute to different aspects, the scheme needs to be integrated from the start with a single public face.

One of the best examples of government and NGO partnership in river conservation is Tweed Forum, a non-profit-making company with charitable status formed in 1991 (www.tweedforum.com). Tweed

Forum is a loose association of 29 different bodies with a shared interest in an integrated approach to managing the catchment of the River Tweed. These bodies include government departments such as Defra, statutory environment and conservation agencies such as SEPA and Natural England, a range of fisheries bodies, and NGOs such as RSPB. The Tweed is the second largest river basin in Scotland, with a total catchment area of 5000 km^2. However, its tributaries flow both from England and from Scotland, and part of the main stem forms the border between the two countries. The river's geographical location brings with it some difficult political challenges, owing to the differences in legislation and government north and south of the border; yet, despite these potential problems, Tweed Forum has been remarkably successful in undertaking a wide range of practical work to improve the management of the catchment. Initiatives have included the Tweed Rivers Heritage Project, which aims to conserve, enhance and raise awareness of the natural, built and cultural heritage, and an invasive species project with the objective of long-term sustainable control of giant hogweed (*Heracleum mantegazzianum*) and Japanese knotweed (*Fallopia japonica*).

The role of non-governmental conservation organizations and the voluntary sector

NGOs in the USA have played a central role in the development of freshwater conservation for natural values, ranging from developing conceptual paradigms to both collaborating with and litigating against government agencies. While NGOs are important in environmental conservation in the UK, they do not appear to play the central role in freshwater conservation that they do in the USA.

In both the USA and the UK, fishermen and angling bodies have played pioneering roles in freshwater conservation. For example, the Anglers Co-operative Association was formed in the UK in 1948. A major goal of the association was to mount legal challenges against pollution of rivers, with one million anglers involved in a grass-roots movement. Improvements, particularly in industrial and sewage discharges, resulted. Similarly, in the USA, the earliest conservation groups with an interest in rivers were primarily fishing organizations. They would seek advice or information from aquatic scientists who were members of the organization or perhaps solicit help from local experts on the design of specific conservation projects (Dewberry & Pringle, 1994).

The UK has a long tradition of voluntary natural history organizations, and its wildlife is the most surveyed of any country in the world. However, rivers and lakes have been relatively ignored by voluntary conservation bodies, perhaps because access is difficult, plants and animals are in and under water, and the sites are mainly on private land. The RSPB is one of the largest voluntary bodies in the UK, with over 1 million members. It has been active for over 50 years in managing and seeking protection for wetlands and waterfowl, but not specifically for protection of lakes and rivers. It is only in the last 10 years that specialist voluntary bodies have been created to protect rivers, such as the River Restoration Centre and the Association of Rivers Trusts (ARTs), which have programmes to promote physical restoration of river channels and to reduce diffuse pollution. However, these are small organizations with memberships in the hundreds. Although the conservation of lakes has not been taken up by the voluntary sector, Pond Conservation promotes the cause of small lakes (typically <1 ha) in terms of surveys, monitoring, conservation management and pond restoration. Pond Conservation, like the two river organizations, relies heavily on financial and organizational support from government agencies.

NGOs in the USA, such as American Rivers, the Audubon Society, the Nature Conservancy, Pacific Rivers Council, the Sierra Club, the Wilderness Society and the World Wildlife Fund, have emerged as key players in freshwater conservation (Pringle & Aumen, 1993; Pringle *et al.*, 1993; Dewberry & Pringle, 1994). NGOs are relying increasingly on science to guide their activities. Almost all national NGOs have senior-level scientists on their board of directors. Many have in-house scientific staff that liaise with the wider scientific community and may also contract out specific projects to consultants and also develop partnerships with academic institutions.

One of the factors that mobilized many US environmental NGOs towards a greater involvement in river conservation was the realization that the National Wild and Scenic Rivers Act could not by itself accomplish the goal of protecting rivers. The first national river organization, American Rivers, was actually founded in 1973 to get stream reaches designated as part of the US Wild and Scenic Rivers System. While this organization initially focused on issues such as 'management versus protection' and 'recreation versus development', these eventually gave way to an emphasis on environmental regulation and pollution control (Gottlieb, 1993). American Rivers and the Sierra Club have since become involved in efforts to reauthorize the Clean Water Act, the Endangered Species Act and in the re-licensing of dams (Coyle, 1993; Woody, 1993). Eventually the focus shifted to the biointegrity and health of river ecosystems (Doppelt *et al.*, 1993).

An example of the key role that NGOs have played in freshwater conservation in the USA is the development, by TNC, of natural heritage programmes (or conservation data centres) which eventually evolved into an international network of biological inventories. These natural heritage programmes collect and manage detailed local information on plants, animals and ecosystems and work to develop information products, data management tools and conservation services to help meet local, national and global conservation needs. The scientific information provided by NatureServe is used by all sectors of society including conservation groups, government agencies, corporations, academia and the public. TNC's Natural Heritage Program initially began for the purpose of documenting the location of rare species at the state level and TNC helped establish the first state natural heritage programme in 1974. This expanded into an international network of biological inventories (NatureServe) which operate in all 50 US states and also in Canada, Latin America and the Caribbean. The NatureServe Network now includes 74 independent natural heritage programmes and conservation data centres containing more than 800 scientists with an annual budget greater than $45 million (www.natureserve.org).

In the UK a similar function is performed by county biological records centres, whose data have been incorporated into national distribution maps for many groups of plants and animals. These centres and other sources of data are now being coordinated through a National Biodiversity Network trust, whose database also contains global biodiversity information (www.nbn.org).

In contrast to the USA, action groups for rivers and lakes are a recent phenomenon in the UK and operate exclusively at the local level, adopting a particular water body or stretch of river. There are currently two lake restoration groups, and both are led by statutory bodies or major NGOs, but encourage participation by local people – the Loe Pool Forum in Cornwall and the Bassenthwaite Restoration Project in the Cumbrian Lake District. There are many more groups involved with rivers. 'River Care' is a loose association of local groups set up in the east of England by the Environment Agency and Anglian Water company, with support from EnCams, the anti-litter charity. They typically form work parties to remove litter from their local stretch of river and develop awareness of its wildlife. A group that dates back to 1990 is Action for the River Kennet (ARK) set up to promote the conservation of this 70 km tributary of the

River Thames. It is strongly opposed to further abstraction by the water company, and promotes physical restoration of the river channel. ARK produces a 'state of the river' report four times each year. The Living River (www.livingriver.org.uk) is a project on the River Avon in the south of England. It is funded by the Heritage Lottery Fund and managed by a partnership of local bodies led by Natural England. It aims to restore sites along the river that the public can visit and enjoy; to provide information about the river for the local communities that live there; and to offer training opportunities to help people understand their relationship with the river system and take responsibility for it.

In 2002, residents from five villages on the borders of Norfolk and Suffolk established the Little Ouse Headwaters Project (LOHP). Its aims are to recreate and maintain a continuous corridor of wildlife habitat along the river valley and to improve its recreation, amenity and education value for the community. LOHP now manages 36 ha of land, including Betty's Fen, which it purchased in 2003. The Fen was degraded by agricultural pollution and drainage. LOHP is undertaking restoration by clearing scrub, recreating open water and mowing the fen margins on a rotational basis. In addition to land management, LOHP organizes guided walks, cycle rides, biological recording and talks by local wildlife experts. In 2006, LOHP was named winner of the national Living Wetlands Award sponsored by RSPB and the Chartered Institution of Water and Environmental Management.

In recent years, the benefits of linking together small independent organizations concerned with river management and conservation have been recognized – for example through the formation of the ARTs in England and Wales. This aims 'to co-ordinate, represent and develop the aims and interests of the member Trusts in the promotion of sustainable, holistic and integrated catchment management and sound environmental practices, recognizing the wider economic benefits for local communities and the value of education' (www.associationofrivertrusts.org.uk). ART has a close association with the newly established Rivers and Fisheries Trusts Scotland (RAFTS), a body whose principal focus is on Scottish salmon fisheries.

NGOs in the USA play the important role of drawing public attention to water resource issues. In 2001, the Audubon Society announced that the US National Wildlife Refuge System was in a state of crisis and was not adequately protecting bird species that are federally listed as threatened or endangered. To emphasize this crisis, the Society released its list of 'Ten Wildlife Refuges in Crisis' (Audubon Society, 2001) because of imminent threats to their biological integrity. Similarly, the US National Parks Conservation Association (NPCA; also an NGO) recently began issuing an annual list of 'America's Ten Most Endangered Parks' to draw attention to the environmental problems (including many water resource issues) facing many park units (NPCA, 2002).

The role of NGOs as watchdogs of federal agencies

NGOs in the USA are playing an increasing role as watchdogs of state and federal regulatory agencies. Many lawsuits are currently under way which involve NGOs as environmental plaintiffs suing federal organizations, such as the US EPA, and states for inadequate water quality compliance with the US Clean Water Act – i.e. failing to implement minimum stream flows where appropriate in the context of allocating pollutant loads within water-quality limited streams. While the watchdog role of NGOs in the USA has resulted in some important advances for freshwater conservation, it can also have negative consequences. Federal land management agencies such as the US Department of Agriculture Forest Service are involved in so many water rights conflicts and lawsuits that it can interfere with the agency's ability to meet its core goals.

A good example of how NGOs serve as watchdogs of federal agencies in the USA is when the Sierra Club sued the US EPA for its failure to enforce the Clean Water Act. The US Clean Water Act requires states to allocate pollutant loads to remedy streams that do not meet water quality standards, through the use of point-source discharge permits. These water quality provisions were ignored until the Sierra Club successfully sued the US EPA (and the state of Georgia's Environmental Protection Division (EPD)). The court ruled in favour of compliance and gave the state of Georgia 5 years to get its Total Maximum Daily Load (TMDL) Program organized. This ruling mandated that Georgia's EPD had to develop TMDLs which quantify the total pollutant load a particular stream segment can handle, and still meet the stream standards. In this instance, the Federal Court enforced provisions of the Clean Water Act by requiring EPA, in the absence of adequate state action, to implement TMDL provisions. Similar lawsuits have been filed in other states within the USA.

The aforementioned events have played a major role in shifting the focus in the USA from controlling point sources of water pollution (i.e. discharges of sewage effluent and industrial water which are controlled through discharge permits under the 1972 Clean Water Act) to addressing also non-point-source pollution, mainly caused by urban and agricultural runoff, drainage of groundwater into surface water, and atmospheric fallout. The main pollutants of concern are nutrients and sediment, but TMDLs could also be expanded to include pesticides, pharmaceuticals and other chemicals of emerging concern. Cost estimates for implementing the TMDL programme range from $900 million to $4.3 billion per year (Gray, 2001), which would be borne primarily by dischargers. While it remains to be seen if desired water quality can be attained in this manner, there is considerable momentum in the USA to move ahead with the programme, by practising adaptive management to make adjustments when necessary (Christen, 2001; National Research Council, 2001).

In summary, the TMDL concept is a radical change from effluent-based standards to ambient water standards and from controlling point sources to controlling entire catchments (Bouwer, 2002). The galvanizing force behind this dramatic shift came from the actions of NGOs. While the TMDL approach was included in the 1972 Clean Water Act, it was largely overlooked until a national regulatory agency was sued and pressured from many environmental groups to implement the programme.

NGOs in the UK have been instrumental in holding the UK government accountable for implementing European directives; however, in contrast to the USA there is little litigation involving freshwater conservation issues that are stimulated by NGOs. This has been possible, however, from 2008 under the new European Directive on Environmental Liability, where interested organizations can require investigations by the statutory conservation agencies into cases of damage to biodiversity and to the water environment. The voluntary bodies are actively engaged in political lobbying, and many – such as the RSPB – have their own parliamentary staff. They can stimulate parliamentary questions to government ministers by briefing Members of Parliament. The major voluntary conservation bodies also maintain a joint office in Brussels – the European Environment Bureau. They have the ear of staff in the EC, and are often included on working parties drafting new legislation or guidance on directives. By contrast, the official conservation agencies are answerable to government ministers and are expected to be politically neutral.

Conclusions

In summary, as pressures on freshwater resources mount, with increasing human population and changing human needs, freshwater conservation objectives are being achieved in the USA and the UK through complex (and sometimes contentious) interactions between many bodies. While the UK has been highly influenced in the last 20 years by international legislation emanating from the EU, the USA has been less influenced by international agreements with respect to overall environmental regulations and to freshwater conservation. Government agencies involved in freshwater resource issues in both the USA and the UK are often faced with a duality of mandates (perhaps more so in the USA), which raises issues of conflict between economic development and environmental protection. Both countries have increasingly taken the approach of forging partnerships, between government organizations and a range of stakeholders, in order to achieve freshwater conservation objectives. In the USA, NGOs have played a key role in the development of freshwater conservation for natural values, ranging from developing conceptual paradigms to both collaborating with and litigating against government agencies. While NGOs are important in environmental conservation in the UK, they do not play the central role in the conservation of fresh waters that they do in the USA.

References

Audubon Society (2001). Refuges in crisis (www.audubon.org/campaign/refuge).

Baron, J. S., Poff, N. L., Angermeier, P. L. *et al.* (2003). Sustaining healthy freshwater ecosystems. *Issues in Ecology*, **10**, 1–16.

Bouwer, H. (2002). Integrated water management for the 21st century: problems and solutions. *Journal of Irrigation and Drainage Engineering*, **128**, 193–202.

Cech, T. V. (2005). *Principles of Water Resources: History, Development, Management and Policy*, 2nd edn. New York: John Wiley.

Christen, K. (2001). TMDL program broken but fixable, NRC report finds. *Water Environmental Technology*, **13**, 31–6.

Coyle, K. J. (1993). The new advocacy for aquatic species conservation. *Journal of the North American Benthological Society*, **12**, 185–8.

Defra (2008). *Future Water: The Government's Water Strategy for England*. London: Department for Environment, Food and Rural Affairs.

Defra, Scottish Executive, Welsh Assembly Government & Department of the Environment Northern Ireland (2005). *Water Framework Directive: Note from the UK Administrations on the Next Steps of Characterization*. London, Edinburgh, Cardiff and Belfast.

Dewberry, T. C. & Pringle, C. (1994). Lotic conservation and science: moving towards common ground to protect our stream resources. *Journal of the North American Benthological Society*, **13**, 399–404.

Doppelt, B., Scurlock, M., Frissell, C. & Karr, J. (1993). *Entering the Watershed*. Washington, DC: Island Press.

English Nature (2006). Target 2010: the condition of England's Sites of Special Scientific Interest in 2005. Peterborough.

Gottlieb, R. (1993). *Forcing the Spring: The Transformation of the American Environmental Movement*. Washington, DC: Island Press.

Gray, R. (2001). EPA sets cost estimate on TMDLs. *Water Engineering and Management*, **148**, 8.

National Parks Conservation Association (NPCA) (2002). *America's Ten Most Endangered National Parks* (www.npca.org/across_the_nation/ten_most_endangered/).

National Research Council (NAS) (2001). *Assessing the TMDL Approach to Water Quality Management*. Washington, DC: National Academy Press.

Pringle, C. M. & Aumen, N. G. (1993). Current efforts in freshwater conservation. *Journal of the North American Benthological Society*, **12**, 174–6.

Pringle, C. M., Rabeni, C. F., Benke, A. & Aumen, N. G. (1993). The role of aquatic science in freshwater conservation: cooperation between the North American Benthological Society and organizations for conservation and resource management. *Journal of the North American Benthological Society*, **12**, 177–84.

Snow, J. (1936). *Snow on Cholera*. London: Oxford University Press; New York: The Commonwealth Fund.

Waxman Report (2004). A special interest case study: the chemical industry, the Bush Administration and European efforts to regulate chemicals. United States House of Representatives Committee on Government Reform – Minority Staff Special Investigations Division, 1 April 2004 (www.democrats.reform.house.gov/Documents/20040817125807-75305.pdf).

Woody, T. (1993). Grassroots in action: the Sierra Club's role in the campaign to restore the Kissimmee River. *Journal of the North American Benthological Society*, **12**, 201–5.

4 · *So much to do, so little time: identifying priorities for freshwater biodiversity conservation in the USA and Britain*

JONATHAN HIGGINS AND
CATHERINE DUIGAN

Introduction

Ecologically unsustainable water- and land-use management practices, pollution, over-harvesting, habitat destruction, invasive species and climate change are major impacts on freshwater ecosystems. As a result, freshwater ecosystems have lost a greater proportion of their species and habitats than ecosystems on land or in the oceans, and they face increasing impacts from these threats (Allan & Flecker, 1993; World Resources Institute *et al.*, 2000; Millennium Ecosystem Assessment, 2005).

Despite, or perhaps because of, this situation, freshwater biodiversity has received relatively little conservation attention and lacks adequate protection. In addition to a lack of attention, there are several key challenges facing freshwater biodiversity conservation. Data on freshwater biota are generally limited, and our understanding of freshwater biota is dependent on sub-sampling environments that we cannot directly view. Freshwater ecosystems are spatially dynamic environments on time frames of days, seasons and years. Freshwater ecosystems have been subjected to significant impacts for centuries, in some cases millennia, and are continuing to be influenced by ever-growing human needs. Strategies to conserve freshwater biodiversity often demand applications at scales and attention to human needs that have not been commonly addressed in terrestrial conservation. These situations are challenging conservation organizations, resource managers and governments to develop new data and approaches to evaluate freshwater biodiversity and to explore strategies unique to freshwater ecosystems.

Given the crisis facing freshwater biodiversity and the limited time and resources to address it, priorities for conservation action need to be focused and appropriately sequenced to achieve success. Recently, there has been

Assessing the Conservation Value of Fresh Waters, ed. Philip J. Boon and Catherine M. Pringle. Published by Cambridge University Press.

a variety of conservation planning approaches developed for freshwater conservation. Many of these approaches have been applied in the USA and Britain. This chapter will provide an overview of several of these approaches that have been applied across large geographical areas to illustrate their rationale and results, strengths and limitations and future directions to improve freshwater conservation planning in the USA, Britain and elsewhere in the world.

Identifying conservation priorities in the USA

The USA harbours a significant proportion of global freshwater biodiversity (Table 4.1). This diversity is in great part due to the size and environmental diversity of the country, which ranges from arctic to subtropical latitudes, includes the Atlantic and Pacific drainages as evolutionary sources of fish, and a tremendous variety of freshwater ecosystems in settings ranging from mountains to grasslands and rain forests to deserts. The USA has more than 5.6 million kilometres (3.5 million miles) of rivers and streams, including the Mississippi/Missouri, which is one of the largest river networks in the world, and tens of thousands of lakes, including the Laurentian Great Lakes shared with Canada, containing 20% of the surface freshwater of the world.

Freshwater ecosystems throughout the USA have suffered impacts from a broad range of sources which have resulted in a significant proportion of species becoming extinct, imperilled and vulnerable (Table 4.1). While these sources of impacts are broad and widespread, altered sediment loads

Table 4.1 *Summary statistics for species of US freshwater fish, crayfish and mussels*

Taxonomic group	Number of described US species	Number of described species worldwide	Percentage of known species worldwide found in the USA	US ranking worldwide in species diversity	Percentage presumed extinct, imperilled or vulnerable in the USA
Fish	799	8 400	10	7	37
Crayfish	322	525	61	1	51
Mussels	292	1 000	29	1	69

Note: Several numbers have been rounded off.
Source: Stein *et al.* (2000).

and non-point-source pollution are major causes of species imperilment in the eastern USA, while exotic species, habitat removal/damage and hydrologic alteration predominate in the West (Richter *et al.*, 1997).

Priority species

Species at risk of extinction are the major components of most conservation planning approaches. In the USA, there are several sources that provide lists of species at risk. The conservation status of species is tracked within every state by a Natural Heritage Program. NatureServe is a non-governmental organization (NGO) that provides standards to Natural Heritage Programs for ranking the global, national and sub-national (state) conservation status of species (www.natureserve.org). These rankings – critically imperilled, imperilled, vulnerable, apparently secure, and secure – are based on species abundance and distribution criteria. All Natural Heritage Programs report on global ranks, and many report on state rankings, depending on state mandates, and they provide locations and estimated viability rankings of species 'element occurrences' or populations. Additional categories used by many federal and state agencies and conservation NGOs include threatened and endangered, imperilled, rare, rapidly declining and endemic. The American Fisheries Society's Endangered Species Committee has recently listed 39% of the species of North American freshwater and diadromous fish as imperilled (Jelks *et al.*, 2008).

The formal listing of threatened and endangered species under the federal Endangered Species Act (see Chapter 7) offers legal standing and protection. Significant numbers of US freshwater mussel (70), fish (138) and crustacean (22) species and sub-species are listed under this Act, in addition to species in many other taxonomic groups (www.fws.gov/endangered/wildlife.html#Species).

Keystone species have a disproportionate impact on the ecological structure and functioning of communities and ecosystems (Paine, 1966). While this concept has been poorly defined and broadly applied (Groves, 2003), the key ecological roles that certain freshwater species play have been well documented. Several migratory fish such as Pacific salmon are known for their dominant and critical roles in riverine systems as well as their impacts on terrestrial systems (Gende *et al.*, 2002).

Umbrella species are those whose conservation confers protection to other co-occurring species (Fleishman *et al.*, 2000). Hunter (2001) suggested that those which have large geographic and home ranges and are habitat generalists make the best examples. Salmon species have been used extensively in the western USA as umbrella species because they are migratory and wide ranging, and individuals cover large expanses of marine and freshwater habitats in their life histories. In the eastern and central USA, several migratory fish species are being used as umbrella species, including the American shad (*Alosa sapidissima*), and the American eel (*Anquilla rostrata*) which migrates from the Sargasso Sea in the Atlantic Ocean to many eastern coastal rivers, and travels over 3000 km to Mississippi River tributaries in the central USA.

Unique stocks

Conserving the evolutionary potential of species requires maintaining their genetic diversity. In the Pacific Northwest of North America, the evolutionary ecology of Pacific salmon, steelhead and cutthroat trout have been a focus of conservation planning because of their economic, ecological and cultural importance, and the federal listing of populations as threatened and endangered. The endangered species committee of the American Fisheries Society defined 214 stocks of Pacific salmon (*Oncorhynchus* spp.), steelhead (*Oncorhynchus mykiss*) and coastal cutthroat trout (*Oncorhynchus clarki clarki*) as being at risk of extinction in Washington, Oregon, Idaho and California (Nehlsen *et al.*, 1991). The Forest Ecosystem Management Assessment Team (FEMAT) increased the number of native stocks of these species designated as at risk of extinction in the Pacific Northwest to 314 (FEMAT, 1993).

Evolutionary Significant Units (ESUs) have been defined for many salmon species in the USA. An ESU is a sub-portion of a species that is defined by substantial reproductive isolation from other conspecific units and represents an important component of the evolutionary legacy of the species (Waples, 1991). Other systematic approaches that take into account non-adaptive, allopatric evolution have been proposed as well (Dimmick *et al.*, 1999). The American Fisheries Society's Endangered Species Committee recently listed 228 sub-species and populations of North American freshwater and diadromous fishes as imperiled (Jelks *et al.*, 2008).

Representative components of aquatic biodiversity

Conserving habitats and levels of biological organization above species are necessary to protect and enhance freshwater biodiversity (Moyle & Yoshiyama, 1994; Angermeier & Schlosser, 1995). Groves *et al.* (2002) address the need for including communities, ecosystems and environmental surrogates in conservation planning to include the many unknown and understudied species, and ecological patterns and processes necessary to allow species to be maintained and evolve across environmental gradients.

Building on a broad body of work, Higgins *et al.* (2005) and Sowa *et al.* (2007) developed geographic information system-based, hierarchical approaches and tools for classifying and mapping aquatic ecological systems and their finer-scale component stream segments and lakes for conservation planning. They characterized aquatic ecological systems as hydrologic landscapes of streams and lakes delineated by catchments of similar landform, geology, elevation and drainage network position. The lakes and streams in these catchments are expected to have similar suites of nutrient and energy sources/dynamics, physical habitats, water chemistry, hydrologic regimes and functionally similar biotic assemblages. These ecological systems are classified and mapped within larger spatial units that define regional patterns of climate and zoogeography.

Aquatic communities have not been widely classified in the USA in a consistent manner, but several classifications have been developed and applied in conservation planning at state and regional levels (Pflieger, 1989). Aquatic community classification is a growing field in the USA, and is also being used to develop indices of biotic integrity and to define the biotic components of ecological systems.

Conservation priorities have also included a broad array of ecological elements such as mussel beds, spawning aggregation sites, waterfowl stop-over sites and unique features, which have been used in conjunction with species and ecological systems in conservation planning.

Planning in ecological and environmental contexts

Terrestrial and freshwater conservation planning is often done simultaneously without appropriate context for freshwater biodiversity. At a regional scale, freshwater biodiversity patterns generally correspond to boundaries of catchments. River basins and smaller catchments have commonly been used as freshwater conservation and management planning units by state and federal agencies for decades. Missouri and California were some of the first states to incorporate physiographic patterns with river basins to develop a combined ecological and environmental context for state-wide freshwater conservation planning (Pflieger, 1989; Moyle & Ellison, 1991).

Ecoregions are large areas of the earth's surface that have similarities in faunal and floral composition and are generally classified using criteria that include climate, geology, soils and vegetation cover (Bailey, 1998).

Several ecoregional frameworks have been used extensively during the past decade for terrestrial biodiversity conservation planning (Olson *et al.*, 2001), but only recently has there been an ecoregional framework developed explicitly for freshwater conservation planning.

Abell *et al.* (2000) developed freshwater ecoregions of North America for regional freshwater biodiversity conservation planning. These ecoregions were delineated based on patterns of fish, mussel and crayfish zoogeography, representing broad biogeographic patterns rather than local endemism. Large river basins are subdivided into different ecoregions. The zoogegraphic patterns depicted by these ecoregions are often associated with historic and/or current climatic and geologic patterns. The freshwater ecoregions of North America have been updated as a component of a comprehensive classification and mapping of the freshwater ecoregions of the world (Abell *et al.* 2008).

Abell *et al.* (2002), Higgins (2003) and Sowa *et al.* (2007) provide approaches to identify the places necessary to conserve the freshwater biodiversity representative of a given region of analysis. Common themes among the guidance are: use ecologically meaningful units for the assessment (freshwater ecoregions, river or lake catchments or other zoogeographic units within political units); use aquatic ecological systems or other surrogates to capture environmental gradients and broad biodiversity representation; use a suite of focal species (those that are rare and endangered, endemic, keystone and wide-ranging); set conservation goals for how much of each biodiversity element needs to be conserved; assess the viability/integrity of and threats to biodiversity elements; and develop a spatially explicit network of conservation areas to represent a vision of conservation success.

Viability and integrity are often evaluated using a combination of information on attributes of size, condition and landscape context (Stein & Davis, 2000). Size is relevant to the abundance, density or the spatial extent of a freshwater species population. Condition is the quality of biotic and abiotic factors, structures and processes within a population or ecological system, such as age structure, species composition, ecological processes and physical/chemical factors. Indicators of the latter categories are often available and include presence of harmful exotic species and degrees of natural flow, nutrient, sediment and temperature regime alteration. Landscape context refers to the quality of structures, processes and biotic/abiotic factors of the landscape surrounding a population or ecological system, including degrees of connectivity and isolation to adjacent habitats, other populations and ecological systems. Spatial data that are commonly evaluated include locations and densities of dams and road/stream crossings, patterns of land use/cover and locations and sizes of urban areas.

A useful indicator of overall integrity is the Index of Biotic Integrity, a concept established by Karr (1981) which uses fish assemblages to evaluate freshwater system integrity. This concept has been further developed and is being implemented broadly across the USA (Simon, 1998). These evaluations assist in highlighting the best examples of biodiversity as priorities for conservation.

Reference conditions are used when possible as benchmarks to characterize many of these attributes. State and federal agencies responsible for monitoring freshwater systems define reference conditions to characterize relative levels of water quality and indices of biological integrity (Davis & Simon, 1995; Simon, 1998).

Spatial priorities

Abell *et al.* (2000) sequenced priorities for conservation planning and actions within ecoregions, providing priorities for the USA from a continental perspective. Ecoregions were ranked using biological distinctiveness (global, continental, bioregional and national levels of species richness and endemism, rare ecological or evolutionary phenomena and rare habitats) and current conservation status (endangered, vulnerable, critical, relatively stable or relatively intact). Those ecoregions that were endangered or vulnerable, and were globally biologically distinct ranked the highest (Figure 4.1). Ecoregions were further ranked using future threats forecasts (future conservation status) to provide an additional perspective on priorities.

These ecoregional priorities guided funding to develop conservation plans within the Tennessee/Cumberland, Mobile Bay, Mississippi Embayment and South Atlantic ecoregions (Smith *et al.*, 2002). Almost all of the geographical areas within other freshwater ecoregions in the USA have subsequently had conservation priorities defined within them, using either these ecoregions or others as assessment units.

Master *et al.* (1998) identified priorities across the conterminous USA to conserve the 307 species of freshwater fish and 158 species of mussels that were classified as imperilled and vulnerable based on the global, national and state conservation status rankings. The spatial analytical framework was the set of 2111 eight-digit hydrologic catalogue units (referred to as 8-HUCs) defined by the US Geological Survey (Seaber *et al.*, 1987). The analysis provided a priority set of 327 8-digit HUCs (15%) that would collectively and efficiently conserve at least one population of each vulnerable and imperilled species of fish and mussels addressed in their analysis (Figure 4.2).

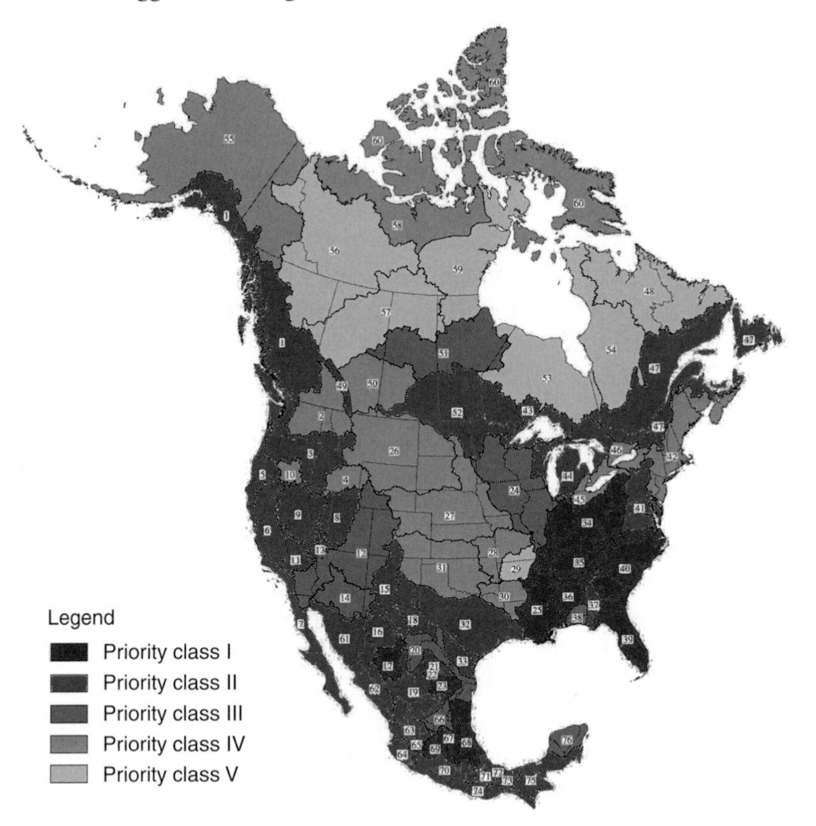

Figure 4.1. Priority classes of freshwater ecoregions of North America based on biological distinctiveness and current conservation status. Source: Abell *et al.* (2000). Reproduced by permission from Island Press, Washington, DC.

There are four species of salmon and steelhead and many of their subspecies/ populations (ESUs) that are listed under the federal Endangered Species Act. Habitats in many areas have been designated for protection. The US National Oceanic and Atmospheric Administration (NOAA) Fisheries program has recently designated critical habitat for 19 salmon and steelhead ESUs under the Endangered Species Act in the states of Washington, Idaho, Oregon and California. They identified more than 48 000 km of stream and shoreline currently inhabited by the targeted salmon stocks, and determined 89% of the habitat as critical (www.nwr.noaa.gov/1salmon/salmesa/crithab/CH-FACT.PDF).

In 2001, the US Congress created the State Wildlife Grants program to fund on-the-ground projects in every state to conserve wildlife and

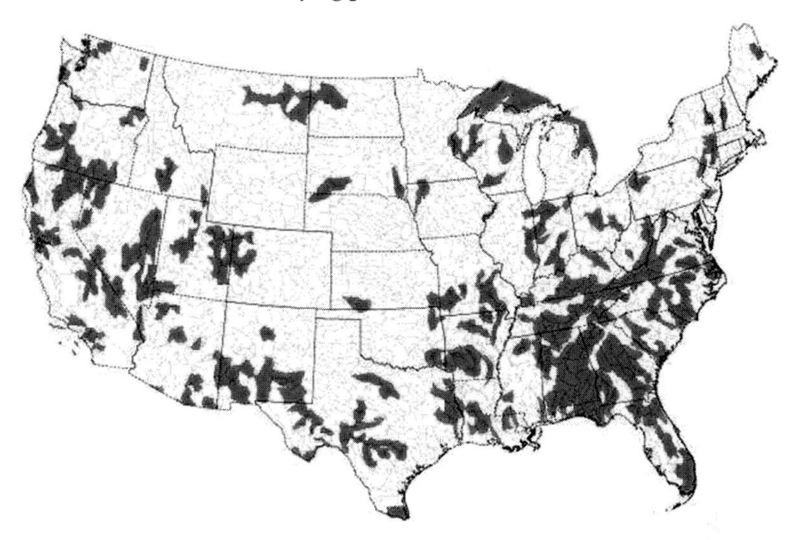

Figure 4.2. Priority set of 327, 8-digit Hydrologic Catalogue Units that contain at least one population of each imperilled and vulnerable species of fish and mussel that occurs in the conterminous 48 United States. Source: Master *et al.* (1998).© The Nature Conservancy and NatureServe.

habitat. To receive congressional funding, each state was required to develop a comprehensive wildlife action plan. These include the following:

- information on the distribution and abundance of species of wildlife including those with low and declining populations and those that are indicative of the diversity and health of the state's wildlife;
- strategies based on the locations and relative conditions of key habitats and community types essential to conserve the assessed species;
- assessment of problems (threats) to species and their habitats;
- descriptions and priorities for conservation actions;
- survey requirements and monitoring plans for species and their habitats and for evaluating the effectiveness of conservation actions and adapting them;
- procedures to review the strategies at intervals not to exceed 10 years;
- coordination of the development, implementation, review and revision of plans.

Additional guidelines from the International Association of Fish and Wildlife Agencies suggested adding landscape, ecosystem and habitat approaches as well as keystone and indicator species, guilds and species of special concern. Other guidelines suggested having explicit goals, maps of focal conservation areas as well as organizational, funding and policy connections (Teaming with Wildlife, www.wildlifeactionplans.org).

All the 50 states and 6 US territories generated plans. These plans generally did a good job of identifying species of concern, and gathering information and documenting threats to species and habitats. However, most plans failed to map focal areas for actions, to set priorities for action or to describe clear and measurable goals for biodiversity or conservation action (Lerner *et al.*, 2006). The state plans vary in their level of attention to freshwater biota. The freshwater components of the plans have been aggregated and made available on the internet (USFWS, 2007).

As an example of the approach to develop a vision for conservation success within a regional, multi-state ecological context, Weitzell *et al.* (2003) conducted a freshwater biodiversity assessment of the Upper Mississippi River basin, which intersects eight states in the north-central USA. This assessment was based on 131 species of fish, mussels, crayfish, other invertebrates (snails and arthropods), reptiles and amphibians that had spatial data available, and 36 ecological system types mapped comprehensively and stratified across basin sub-regions. They identified a suite of areas of freshwater biodiversity significance and connectivity corridors that comprised at least one population of 102 (78%) of the species targets (45% meeting overall distribution and abundance goals) and at least one example of each ecosystem type stratified across the 21 basin sub-units, meeting all of their distribution and abundance goals (Figure 4.3a). Furthermore, freshwater priorities were combined with existing terrestrial conservation priorities to create a map of 47 high-priority conservation areas throughout the basin (Figure 4.3b).

Numerous individual regional plans such as that of Weitzell *et al.* (2003), developed by The Nature Conservancy and partners, have identified over 4000 areas of biodiversity significance across the USA. Other assessments, updates, refinements and applications of new and more detailed approaches are continuing across the USA. A next useful step will be to incorporate all of the freshwater conservation priorities from the variety of assessments and plans into a single, national database. The spatial priorities and information available for freshwater biodiversity conservation generated from these assessments are being used to inform the next phases of site-specific, multi-site and basin-wide strategies. These assessments are not meant to be a single solution over time. Information and priorities will change in an adaptive manner in response to changes in biodiversity status, threats and conservation management and through implementing better and more refined approaches and tools to evaluating and planning for biodiversity conservation.

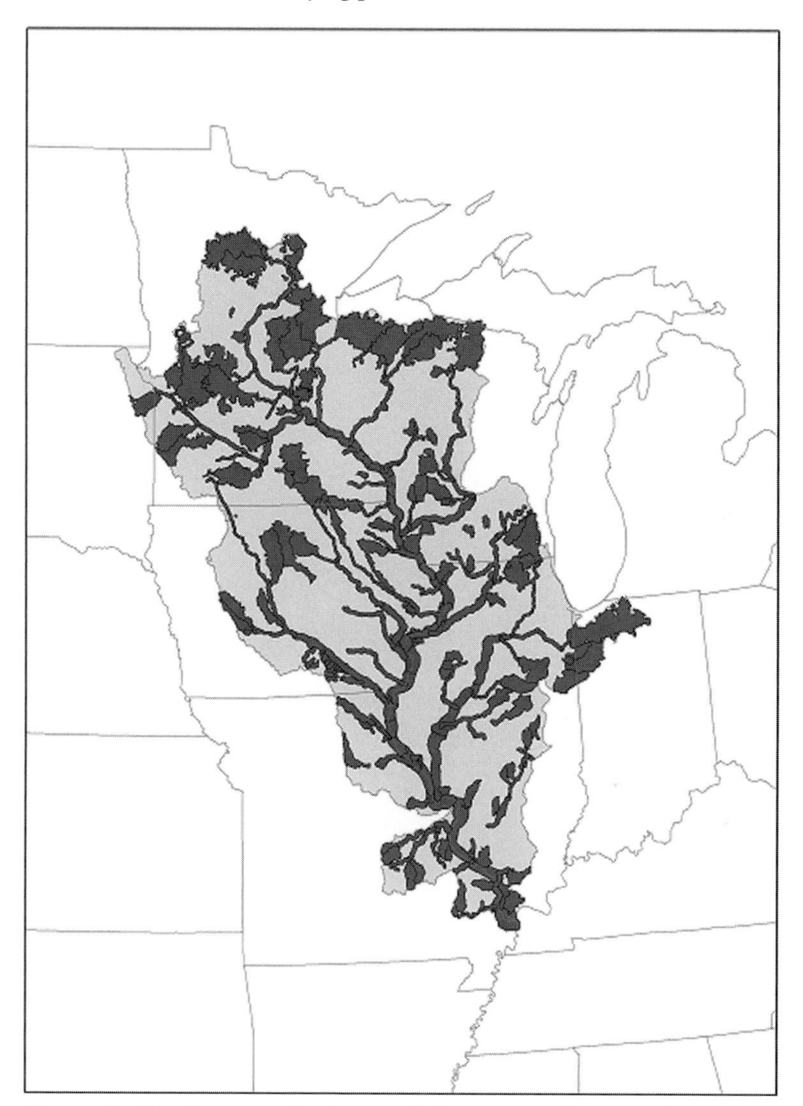

Figure 4.3a. Priority areas of freshwater biodiversity significance and connectivity corridors in the Upper Mississippi River basin. Source: Weitzell *et al.* (2003).© The Nature Conservancy and NatureServe.

Identifying conservation priorities in Britain

Located on the western margins of the European continent and Scandinavia, the islands of Britain support unique biodiversity resources that are a product of their post-glacial history, climate, geographical

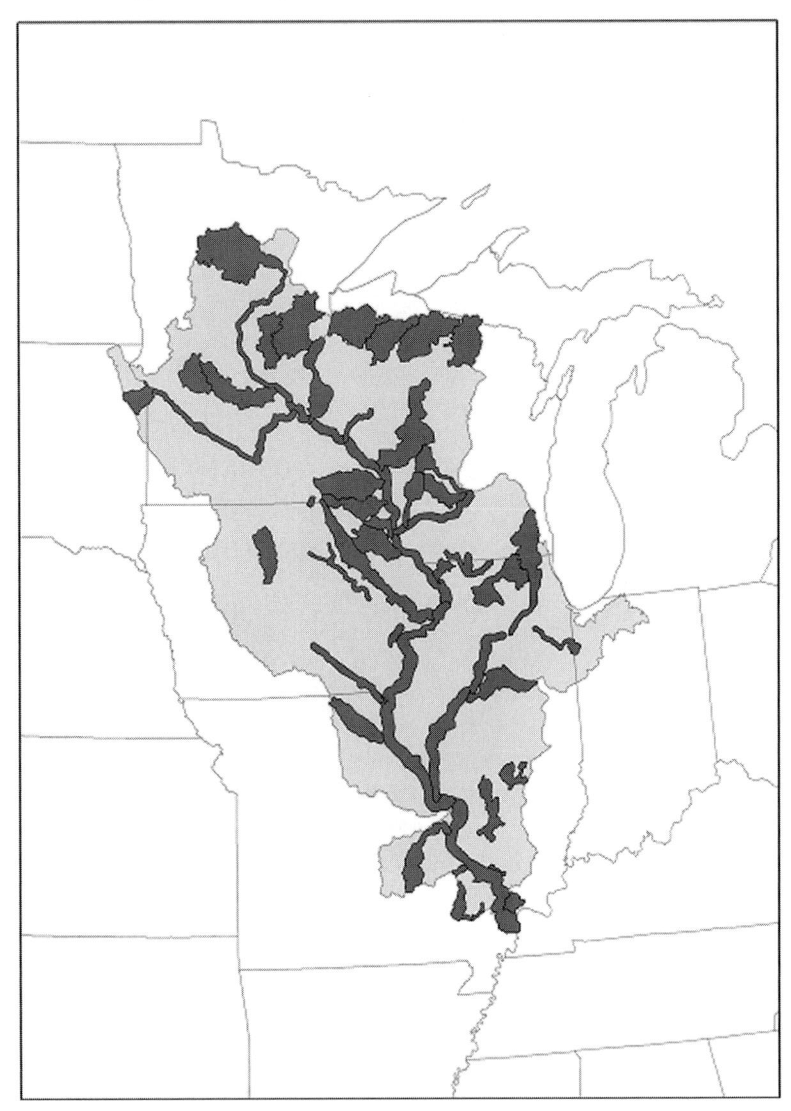

Figure 4.3b. Priority areas of combined freshwater and terrestrial biodiversity significance and connectivity in the Upper Mississippi River basin. Source: Weitzell *et al.* (2003).© The Nature Conservancy and NatureServe.

location, ecological conditions and colonization history. The total surface area of standing water in Britain is 213 911 ha (i.e. 1% of the land surface) consisting of 43 738 individual water bodies (Hughes *et al.*, 2004). Most water bodies have an area smaller than 1 ha with less than 10% having a surface area larger than 10 ha. The largest freshwater lakes in Britain are found in Scotland and include Loch Lomond (7073 ha; maximum depth 190 m) and Loch Morar (2683 ha; maximum depth 310 m). In a global context, they are amongst the deepest recorded lakes. It has also been estimated that the number of small water bodies in Britain exceeds 100 000, but their accumulated area is probably not more than 5% of the total for Britain, with their volume less than 1% (Hughes *et al.*, 2004).

Data on linear extent for rivers are more difficult to establish but Smith and Lyle (1979) identified 1445 individual river systems with 7835 associated streams. The island location restricts distance from the sea, resulting in the majority of river systems being less than fifth order. However, there are several larger rivers, such as the River Tay (Scotland) and the Humber (England), which are eighth order.

The freshwaters of Britain and their biota are vulnerable to a diversity of human impacts including water abstraction and transfers, impoundments, pollution (especially acidification and eutrophication), inappropriate catchment management, invasive species, industrial and urban development within catchments, recreational uses and water transfers. Acidification and eutrophication are the most widely occurring threats to water quality. For example, Kernan *et al.* (2004) demonstrated that approximately 31% of lakes (>0.02 ha) are vulnerable to acidification and they were mainly located in the sensitive upland areas of Britain.

Britain supports a wide diversity of running waters, from the torrential acid mountain streams of the north and west, through the slow-flowing lowland rivers of eastern England to the alkaline chalk streams of southern England. Standing waters exhibit a comparable level of diversity from the clear corrie lochs, llynnau and tarns often with relatively sparse vegetation, through to large lowland lake systems dominated by stands of emergent, submerged and floating vegetation. This natural resource is supplemented by the biota which has become established in artificial water bodies, such as farm ponds, drainage ditches, reservoirs and canals. It does not exhibit a comparable continental range of aquatic biodiversity when compared with North America but it does occupy an important position in the Atlantic biogeographical region recognized by the European Union.

Priority habitats

In a conservation context in Britain and the larger United Kingdom, identifying priority freshwater habitats has focused on key habitats perceived to be at risk at a regional and international level and securing a representative series of habitats within protected areas.

A range of freshwater habitats have been identified under the UK Biodiversity Action Plan (BAP) (see Chapter 2) for which Habitat Action Plans (HAPs) have been developed: aquifer-fed naturally fluctuating water bodies, rivers, ponds, eutrophic standing waters and dystrophic, oligotrophic and mesotrophic lakes. Each HAP describes the current status and condition, factors affecting the habitat and actions under way. Action plan objectives and targets covering the management and restoration of the habitat are included. Organizations are identified to provide legislative protection, site safeguard and management, advice, research and monitoring, information and publicity.

For example, the estimated 39 chalk rivers (within the rivers HAP) of south and east England are characterized by their plant communities which are dominated mid-stream by river water crowfoot (*Ranunculus penicillatus* var. *pseudofluitans*) and water starworts (*Callitriche obtusangula* and *Callitriche platycarpa*), with marginal beds of watercress (*Rorippa nasturtium-aquaticum*) and lesser water-parsnip (*Berula erecta*). These streams are largely groundwater fed and have intermittent stretches in their headwaters. The objectives of the plan are to maintain the characteristic biota, and to restore qualifying river sections to what is considered a favourable environmental condition.

In addition, habitat statements have been produced for two broad freshwater habitat categories – 'rivers and streams' and 'standing open water and canals'. This highlights the specific need to consider these habitat types in Local Biodiversity Action Plans (LBAPs).

In 1992, the European Community adopted a Directive with the principal aim 'to contribute towards ensuring biodiversity through the conservation of natural habitats and of wild fauna and flora in the European territory of the Member States to which the Treaty applies' (*Council Directive 92/43/EEC on the conservation of natural habitats and of wild fauna and flora*, commonly known as the Habitats Directive). Table 4.2 lists the freshwater habitats and species of Community interest which occur in Britain and are included in Annex I and Annex II of the Directive, and therefore considered conservation priorities at a European level. These species and habitats are to be protected throughout the European Union by the creation of a series of Special Areas of Conservation (SACs) and by other safeguard measures for particular species. At present, the UK series of SACs comprises 611 sites covering a total area of more than 2 504 662 ha (Joint Nature Conservation Committee – www.jncc.gov.uk).

Table 4.2 *Listing of freshwater habitats and species in Annex I and Annex II of the Habitats Directive which occur in Britain*

Habitats

Oligotrophic waters containing very few minerals of sandy plains (*Littorelletalia uniflorae*)

Oligotrophic to mesotrophic standing waters with vegetation of the *Littorelletea uniflorae* and/or of the *Isoëto-Nanojuncetea*

Hard oligo-mesotrophic waters with benthic vegetation of *Chara* spp.

Natural eutrophic lakes with *Magnopotamion* or *Hydrocharition*-type vegetation

Natural dystrophic lakes and ponds

Mediterranean temporary ponds*

Turloughs*

Water courses of plain to montane levels with the *Ranunculion fluitantis* and *Callitricho-Batrachion* vegetation

Species

Freshwater pearl mussel *Margaritifera margaritifera*

Southern damselfly *Coenagrion mercuriale*

White-clawed (or Atlantic stream) crayfish *Austropotamobius pallipes*

Sea lamprey *Petromyzon marinus*

Brook lamprey *Lampetra planeri*

River lamprey *Lampetra fluviatilis*

Allis shad *Alosa alosa*

Twaite shad *Alosa fallax*

Atlantic salmon *Salmo salar*

Spined loach *Cobitis taenia*

Bullhead *Cottus gobio*

Great crested newt *Triturus cristatus*

Otter *Lutra lutra*

Floating water-plantain *Luronium natans*

Slender naiad *Najas flexilis*

*priority habitat, i.e habitat type in danger of disappearance and whose natural range mainly falls within the territory of the European Union.

These SACs, together with Special Protection Areas (SPAs) designated under the Birds Directive, form the Natura 2000 network extending across Europe.

The CORINE (Coordination of Information on the Environment) habitat classification (European Commission, 2003) provided the basis for interpreting the habitat descriptions at a European level. In order to select SACs at a country level, it was also necessary to interpret these descriptions at a regional level (Table 4.3). In Britain, eight freshwater habitat types

Table 4.3 *Rarity status of aquatic charophytes in Britain*

	WCA Sch	Red List	NR/DD	Nat Sc	Spec Resp	BAP
Chara aculeolata (pedunculata)				NS		
Chara baltica			VU			SAP
Chara canescens	8	EN				SAP
Chara connivens		EN				SAP
Chara curta				NS	SR	SAP
Chara fragifera		VU				
Chara intermedia		EN				
Chara muscosa			DD		SR	SAP
Chara rudis			NR			
Lamprothamnium papulosum	8		NR			
Nitella confervacea			NR			
Nitella flexilis				NS		
Nitella gracilis		VU				SAP
Nitella mucronata				NS		
Nitella tenuissima		EN				SAP
Nitellopsis obtusa		VU				SAP
Tolypella glomerata				NS		
Tolypella intricata		EN				SAP
Tolypella nidifica		EN				SAP
Tolypella prolifera		EN				SAP

WCA Sch – Species included in Schedule 8 of the Wildlife and Countryside Act 1981
Red List – Species included in British Red Lists.
Threat categories as in IUCN Species Survival Commission (2000) and Wigginton (1999): CR – Critically Endangered, EN – Endangered, VU – Vulnerable.
NR/DD - Nationally Rare (NR) species (previously called Near Threatened) are without an IUCN Red List designation in Britain, but have been recorded as native since 1986 in 15 or fewer $10 \times 10\,km^2$ in Britain; Data Deficient (DD) species are those for which there is insufficient information to make an adequate assessment of threat.
Nat Sc – Nationally Scarce species.
These are plants that have been recorded as native since 1986 in 16–100 $10 \times 10\,km^2$ in Britain and are without an IUCN Red List designation. Nationally Scarce vascular plants are listed in Stewart *et al.*, (1994), but a few designations have been changed as a result of more recent data in Preston *et al.* (2002).
Spec Resp – Species that are endemic or near-endemic to Europe and for which Britain has 'special responsibility' because it supports a high proportion (certainly or probably more than 25%) of the European population (Chris Preston, David Pearman, M. Palmer, pers. comm.; Stewart & Church, 1992, for charophytes).
BAP – Priority plant species with a current Species Action Plan under the UK's Biodiversity Action Plan.
Source: Extracted from Duigan *et al.* (2007).

were recognized, including two 'priority' habitats under the Directive (i.e. habitats in danger of disappearance and whose natural ranges mainly fall within the territory of the European Union; see Table 4.2). Some of these habitat types are relatively common and extensive in Britain (e.g. oligotrophic to mesotrophic standing waters with vegetation of the *Littorelletea uniflorae* and/or of the *Isoëto-Nanojuncetea*). However, they represent conservation priorities and can contribute significantly to the maintenance of those habitat types at favourable conservation status from a European perspective.

In contrast, some habitat types have localized distributions. For example, 'hard oligo-mesotrophic waters with benthic vegetation of *Chara* spp.' are defined as water with a high base content (usually calcium) confined to areas of limestone and other base-rich substrates. These lake types are found mainly in three types of situations: lakes on limestone substrate; on coastal deposits of calcium-rich shell-sands (e.g. machair lochs); and lakes with nutrient inputs from other base-rich influences, such as serpentine and boulder clays. Abundant charophytes (stoneworts) are the characteristic component of the vegetation where they can occur as dense stands covering a significant part of the lake bed overlying marl deposits. The representative SACs cover the wide geographical range of the habitat in England, Scotland and Wales and the variety of topographical and ecological situations described. Sites unimpaired by nutrient enrichment and containing a number of rare and local *Chara* species were identified as conservation priorities.

Priority species

In common with the USA, species conservation priorities in Britain have also centred on those perceived to be rare or endangered. The Joint Nature Conservation Committee operates the Species Status Assessment Project which uses IUCN Red Data Book criteria and categories and some additional domestic categories to assign conservation status to the qualifying species. A list of 'Species of Conservation Concern' (SoCC) is thereby produced. In addition, project reviews are published periodically and cover some specific freshwater groups (e.g. stoneworts; Stewart & Church 1992).

The SoCC criteria are: (1) threatened, endemic and other globally threatened species; (2) species for which the UK has more than 25% of the world (or appropriate biogeographical) population; (3) species where numbers or range have declined by more than 25% in the last 25 years; (4) in some instances where the species is found in fewer than $15 \times 10\,km^2$ in the UK; and (5) species listed in the conservation legislation (EC Birds or Habitats Directives (see Table 4.2), the Bern Convention or under the Wildlife and Countryside Act 1981). A number of freshwater species across a range of plant and animal groups qualify as conservation priorities using these criteria.

In addition to UK BAP, Priority Species are those species which are globally threatened or are rapidly declining in Britain (i.e. by more than 50% in the last 25 years). There are also a number of additional domestic level priority categories, such as Nationally Rare Species, Nationally Scarce Species, and Species for which Britain has Special Responsibility. These multiple levels of conservation priority can quickly generate complex matrices for plant or animal groups (Table 4.3). The Species Status Assessment Project seeks to be the up to date and best source of species status information in a world of changing conservation priorities.

Britain has a relatively restricted fish fauna in comparison with the European continent, but the island status of certain fish populations adds to their conservation importance, as they represent unusual or isolated forms, or are internationally important populations in their own right. Eight freshwater fish species found in Britain are listed in Annex 2 of the Habitats Directive (Table 4.2). Some of these species are also the subject of SAPs (e.g. shad, *Alosa alosa* and *Alosa fallax*), with additional taxa thought to be important glacial relics (e.g. the gwyniad, *Coregonus lavaretus*) or isolated populations (e.g. vendace, *Coregonus albula*). Criteria for selecting important sites for freshwater fish have been developed and include the presence of rare or pristine stocks and local races, and unusual or diverse communities (Maitland, 1985). There is scope for the expansion and recognition of the conservation importance of other important fish species and races, such as Arctic charr (*Salvelinus alpinus*) and genetically distinct trout (*Salmo trutta*) populations (Duigan, 2005; Giles, 2005).

Finally, in Britain, salmon are often promoted as indicator species of healthy rivers (Doughty & Gardiner, 2003; Mawle & Milner, 2003). However, they are extinct from some of the major river systems in Europe. British (and especially Scottish) salmon represent a significant proportion of the total European stock. SAC site selection for this species has incorporated the variation in ecological and hydrological characteristics of supporting rivers, and the life-cycle strategies adopted by the resident salmon.

Spatial priorities

The two previous sections described how conservation priorities in Britain are set using habitat- and species-focused criteria at national and international levels. This section describes the geographical prioritization approach which underpins our national conservation site network. The conservation agencies in Britain are legally obliged to select and notify any area which is 'of special interest by reason of any of its flora, fauna, or geological or physiographical features'. These conservation sites are referred to as Sites of Special Scientific Interest (SSSIs). The overall regional objective is to identify the most important conservation areas representative of the diversity of species and habitats found in Britain, 'so that society may use and appreciate its value to the fullest extent' (Nature Conservancy Council, 1984). The majority of these nationally important sites are held in private ownership, but are often subject to management agreements to safeguard the features of conservation interest. Guidelines for site selection present the rationale for site evaluation and selection (Nature Conservancy Council, 1989). The evaluation criteria for habitats were primarily naturalness, fragility, diversity, how typical they are, size (extent) and recorded history, as discussed in the Nature Conservation Review (Ratcliffe, 1977). Supplementary considerations were made for rare species and species assemblages.

It was recognized from the outset that truly natural habitats (i.e. unmodified by human influence) were likely to be relatively rare, so the term 'semi-natural' was coined and applied to 'modified types of vegetation in which the dominant and constant plant species are accepted natives to Britain, and structure of the community conforms to the range of natural types'. Fragility was related to sensitivity to environmental change. Diversity was defined as the numbers of both communities and species, which were closely related to habitat diversity. Typical sites were selected for their characteristic or common habitats, communities and species. The existence of a scientific research record also added to the conservation value of a site. Since then, more detailed scoring and evaluation systems have been developed for individual site assessments, such as SERCON (System for Evaluating Rivers for Conservation) and LACON (see Chapters 7 and 8), but the original NCR criteria have been retained.

Individual sites also need to be placed in the network of conservation areas selected with reference to a regional habitat classification scheme and thereby ensuring there is representative coverage across environmental gradients and habitat transitions. This is dependent on having a detailed knowledge of the extent, distribution and condition of habitat types throughout the region. In Britain, three major classification schemes for standing waters, lowland ditch systems and flowing waters are employed in the selection of SSSIs (Nature Conservancy Council, 1989; Palmer, 1992; Palmer *et al.*, 1992 (now superseded by Duigan *et al.*, 2007); Holmes *et al.*, 1999). These classification schemes are based mainly on aquatic vegetation, but it is also possible to select SSSIs on the basis of invertebrates, birds, amphibians, fish or otters (Nature Conservancy Council, 1989). For example, in the case of standing waters the latest classification system was devised from a TWINSPAN analysis of an extensive dataset of submerged and floating macrophyte records from 3447 lakes in England, Scotland and Wales (Duigan *et al.*, 2007). Separate ecological descriptions of 11 distinct lake groups were produced. British lake and river conservation evaluation schemes are being extended beyond the almost exclusive emphasis on aquatic plants, including habitat attributes (Raven *et al.*, 1998; Rowan *et al.*, 2006), and specific aquatic plant communities within a water body can be placed in a national context with reference to the National Vegetation Classification (Rodwell *et al.*, 1995).

Under the guidelines for SSSI selection (Nature Conservancy Council, 1989), it was determined that the habitat classification schemes should be related to geographical subdivisions known as 'Areas of Search' (AOS) but they are administrative regions largely undefined by topography, geology or land-use patterns. The conservation site series (SSSIs) within each AOS should represent all the different habitats and species that are present by at least one example or population. It was acknowledged from the outset that the subdivision of Britain into biogeographical areas would give the most satisfactory basis for site selection, but there was no agreed system available at the time. More recently, Palmer (1999) advocated the pursuit of freshwater conservation within a natural framework, as distinct from administrative areas. She defined 12 biogeographical zones from the analysis of climate, relief, geology, soils, land use and occurrence data for 300 native freshwater species (macrophytes, dragonflies, aquatic molluscs, amphibians and selected leeches, water beetles and crustaceans) on a $10 \times 10\,km^2$ grid. However, this approach does not have the same rigour as the North American ecosystem approach which used species strictly confined to aquatic environments (i.e. fish, mussels and crayfish). For example, it is evident that some of the zones identified have a marginal coastline distribution independent of potential common catchment characteristics. It has yet to be adopted as part of the conservation site-selection process.

There are some similarities between the North American freshwater ecoregions approach and the biogeographical zones subsequently developed for conservation planning in England and Scotland. The Scottish Natural Heritage Zones and English Natural Areas are geographical units that share similar natural heritage characteristics defined on the basis of information on geology and geomorphology, species and habitats, landscape character and land use, topography and other geographical factors, but their boundaries do not necessarily relate to catchments. In Europe, an early exercise to identify ecoregions for rivers and lakes was based on the fauna living in these waters as presented in Limnofauna Europaea (Illies, 1978). In this study a total of 25 freshwater ecoregions are demarcated across the continent on their own or amalgamated into larger biogeographical regions, with a view to providing a framework that supports European legislation and state of the environment reporting.

More recently, physical freshwater habitat classifications have been produced as part of a typology exercise for the EC Water Framework Directive (WFD). In Britain these lake types are primarily based on geology and mean lake depth (see www.wfduk.org/). For rivers, the types have been identified using altitude and alkalinity. Now there is an opportunity to match available wide-scale biological datasets for freshwater habitats and species with this physical water body framework from the WFD typology exercise at a catchment scale. It is tempting to predict that a number of ecoregions could be recognized from the resultant biodiversity patterns. This would allow conservation site selection to be carried out at a more meaningful scale based on ecological processes. For example, local sites could be selected as representative of catchment biodiversity with catchments being selected as representative of ecoregions.

Assessing ecological condition – conserving the best

In the UK, relatively few water bodies are likely to be in a pristine or reference condition and any of those that are should be considered priorities for conservation. This appreciation of naturalness has always been a fundamental criterion for conservation site assessment and it is becoming increasingly important and subject to theoretical analysis (O'Sullivan, 2005). Although Britain has some of the longest established freshwater datasets in the world, there are still few datasets which are consistently collected and extend back more than a couple of decades. Most of these would have started when a water body could no longer be considered in reference condition. Near-pristine conditions may be defined using historic datasets, modelling and other hind-casting methods and/or palaeolimnology. It may be possible to carry out space-for-time substitutions by using relatively remote ecosystems for undisturbed environmental conditions (Catalan *et al.*, 1993).

Under the WFD, categories of ecological status for a water body are defined as: high, good, moderate, poor or bad, based on their degree of deviation from the reference 'high' status (see Chapter 2; Pollard & Huxham, 1998). This 'high' or near-pristine condition exists where the biological, hydromorphological and chemical quality elements correspond completely, or nearly completely, with undisturbed conditions. This approach recognizes that an assessment of naturalness is also a fundamental principle for assessing the ecological quality of sites at an international level. Ecological descriptions of individual reference conditions for the WFD water body types defined under the WFD can be found at www.wfduk.org/. Having an ecological picture of reference condition greatly assists with management target setting (Anderson, 1993; Smol, 2008).

Discussion

Conservation priorities are identified in the USA and Britain in a variety of ways. Figure 4.5 provides a summary and comparisons of the major criteria used to define them. It is not surprising that both the USA and Britain have rare species and those at risk as conservation priorities. Species with these attributes have long been conservation priorities globally. The USA and Britain both rely on globally ranked species and rankings of species for their political geographies. Britain also uses European Union criteria to define species and habitat priorities, which is a continental geographic perspective, but bounded politically. There is a multitude of endemic species in the USA, and they are almost always included as conservation priorities.

In the USA, keystone and umbrella species are used to confer protection to other components of biodiversity, while in Britain, salmon are largely used as indicators of environmental quality or integrity. There are more wide-ranging species in the USA than in Britain, and the terms above have become more formalized.

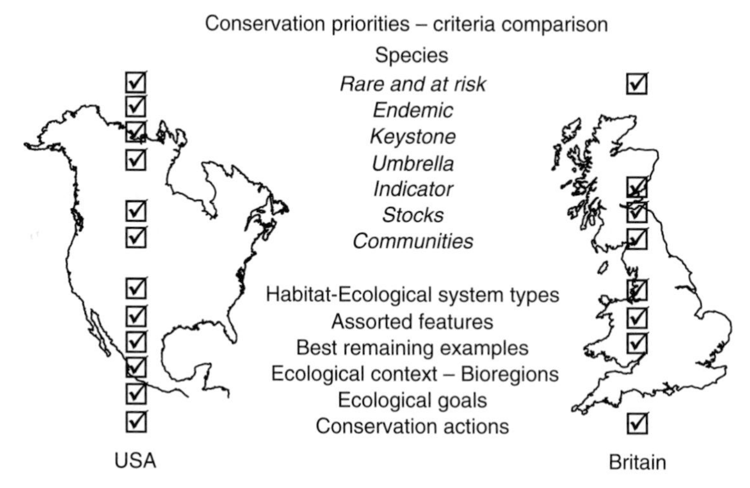

Figure 4.5. Summary of major criteria to define freshwater conservation priorities in the United States and Britain. Many of the differences in the way areas are mapped are due to different approaches to identifying and mapping conservation priorities in each ecoregion. This map was not yet complete at the time of publication, but is being shown to illustrate the extent and degree of freshwater biodiversity planning that has taken place within ecoregional contexts across the conterminous USA. Freshwater conservation priorities have also been identified in Hawaii and Alaska.

Regardless of the terms, conservation priorities identified using species categorized as above in the USA and Britain will more than likely confer protection to other sympatric species.

In both the USA and Britain, genetically distinct strains are identified as priorities, but the level of documented distinction within the array of species is limited. Several approaches attempt to protect evolutionary potential by conserving species across the environmental variability in their ranges, and by conserving isolated populations. There is an ongoing debate among evolutionary ecologists and conservation planners regarding the formal ways evolutionary ecology distinctions are described, mapped and used in conservation planning (Mayden & Wood, 1995). Through additional research on evolutionary ecology and genetics, more exacting priorities for conserving the evolutionary potential of species may be identified.

The broad array of river and lake types, either defined as habitats or components of ecological systems, are conservation priorities in the USA and Britain. In addition, Britain defines those habitats that are at risk of disappearance and which are primarily found in Britain. The simpler classification of habitats in Britain is a product of the limited range of habitats compared with the USA, and of a finer-scale approach to dividing rivers into distinct units in the USA, but this approach is being developed in Britain. Regardless, the goal to conserve the geographically representative habitats is shared. In the USA, this goal is more of a focus of conservation NGOs than of most state and federal agencies but it is a core approach used by government conservation agencies in Britain.

Conservation of the best remaining examples is a priority in Britain and the USA. This approach secures high-quality examples of species and habitats. However, conservation priorities defined only from the perspective of condition and not in conjunction with biodiversity patterns can result in protection of a sub-set of habitats, which for several reasons may be predominantly in good condition, at the expense of the best remaining examples of a broad array of other habitats which face continuing threats.

In the USA, catchments and ecoregions have been used extensively as ecological and environmental contextual planning units. A catchment approach would facilitate the selection of a more meaningful spatial network of sites in Britain (as advocated by Ratcliffe, 1977). In Britain, another effort should be made to produce a map of freshwater biogeographical regions as a basis for future conservation site selection. Because of differences in land scale they may not conform to the North American definition of bioregions but would be operational equivalents. Combining key physical, chemical and biological datasets could produce this map. The WFD may provide an opportunity to do this in the near future through using the physical typologies for rivers and standing waters and the datasets of obligate freshwater organisms to describe regional patterns in the catchment. Regional freshwater biodiversity patterns are starting to emerge as more analyses of wide-ranging datasets are carried out. Using an ecological unit for planning also promotes setting biodiversity conservation goals at an appropriate scale. However, these goals are still testable hypotheses, and require further evaluations and refinements.

Conservation actions are defined both in Britain and the USA, depending on the conservation programme being used as a framework. Many conservation priority sites in the USA have not yet had conservation plans and strategies defined for them. Knight *et al.* (2006) suggest that assessments alone without strategies for action are inadequate. Given the vast number of conservation priorities identified in the USA, the next step after conducting an assessment is to develop and implement strategies. The state wildlife action plans are an attempt to provide strategies in addition to conservation priorities. In Britain, one example of identifying conservation actions is in defining conservation priorities within HAPs and SAPs. The UK LBAPs could be considered operational equivalents to the US state wildlife action plans.

Freshwater biodiversity conservation planning in both the USA and Britain is often done in conjunction with terrestrial planning. A sole focus on freshwater biodiversity can result in a more comprehensive and ecologically meaningful freshwater plan. However, unless freshwater priorities are integrated at some point with those for terrestrial and marine areas, opportunities can be missed to optimize strategies that would be beneficial to freshwater biodiversity. A comprehensive and integrated perspective on freshwater, terrestrial and marine conservation priorities and opportunities will promote more efficient and effective conservation outcomes.

Conservation planning in Britain and the USA is carried out by agencies, academics and conservation organizations. Sometimes these efforts are well coordinated for specific regional or site-planning activities, but they have not often been used to generate comprehensive national plans. A greater level of coordination to share strengths and resources would greatly enhance conservation planning rigour and efficiency.

While the methodological and geographical diversity in approaches makes it difficult to generate a single, consistent perspective for conservation priorities, there is an opportunity to learn from the variety of lessons made available from these approaches. Ultimately, there is no one best way to define priorities. Priorities are based on goals, limitations and the questions being asked. The USA and Britain have collectively taken broad, comprehensive and proactive approaches to setting priorities for conserving the biodiversity representative of their countries. These approaches not only prioritize the last of the least in order to lose fewer species and habitats in future, but also to incorporate the best of the rest to keep more species and habitats from becoming at risk.

References

Abell, R., Thieme, M., Revenga, C. *et al.* (2008). Freshwater ecoregions of the world: a new map of biogeographic units for freshwater biodiversity conservation. *BioScience*, **55**, 403–414.

Abell, R. A., Olson, D. M., Dinerstein, E. *et al.* (2000). *Freshwater Ecoregions of North America: A Conservation Assessment.* Washington, DC, World Wildlife Fund – US: Island Press.

Abell, R. M., Thieme, M., Dinerstein, E. & Olson, D. (2002). *A Sourcebook for Conducting Biological Assessments and Developing Biodiversity Visions for Ecoregion Conservation. Volume II: Freshwater Ecoregions.* Washington, DC, USA: World Wildlife Fund (www.worldwildlife.org/science/pubs/FWsourcebook2002.pdf).

Allan, J. D. & Flecker, A. S. (1993). Biodiversity conservation in running waters. *BioScience*, **43**, 32–43.

Anderson, N. J. (1993). Natural versus anthropogenic change in lakes: the role of the sediment record. *Trends in Ecology and Evolution*, **8**, 356–61.

Angermeier, P. L. & Schlosser, I. J. (1995). Conserving aquatic biodiversity: beyond species and populations. In *Evolution and the Aquatic Ecosystem: Defining Unique Units in Population Conservation*, ed. J. L. Nielsen, American Fisheries Society Symposium 17. Bethesda, MD: American Fisheries Society, pp. 911–27.

Bailey, R. G. (1998). *Ecoregions: The Ecosystem Geography of Oceans and Continents.* New York: Springer-Verlag.

Catalan, J., Ballesteros, E., Gacia, E., Palau, A. & Camarero, L. (1993). Chemical composition of disturbed and undisturbed high-mountain lakes in the Pyrenees: a reference for acidified sites. *Water Research WATRAG*, **27**, 133–41.

Davis, W. S. & Simon, T. P. (eds.) (1995). *Biological Assessment and Criteria: Tools for Water Resource Planning and Decision Making.* Boca Raton, FL: CRC Press, Inc.

Dimmick, W. W., Ghedotti, M. J., Grose, M. J. *et al.* (1999). The importance of systematic biology in defining units of conservation. *Conservation Biology*, **13**, 653–60.

Doughty, R. & Gardiner, R. (2003). The return of salmon to cleaner rivers – a Scottish perspective. In *Salmon on the Edge*, ed. D. Mills. Oxford: Blackwell Science, pp. 175–85.

Duigan, C. A. (2005). Why do we care about upland waters? In *The Future of Britain's Upland Waters*, eds. R. W. Battarbee, C. J. Curtis & H. A. Binney, Proceedings of a meeting at the Environmental Change Research Centre, University College London, 21 April 2004. London: Environmental Change Research Centre, University College London, UCL Environment Institute and Department for Environment, Food and Rural Affairs, pp. 8–11.

Duigan, C. A., Kovach, W. & Palmer, M. (2007). Vegetation communities of British lakes: a revised classification scheme for conservation. *Aquatic Conservation: Marine and Freshwater Ecosystems*, **17**, 147–73.

European Commission (2003). *Interpretation Manual of European Union Habitats.* EUR 25. Brussels: European Commission, DG Environment.

FEMAT (Forest Ecosystem Management Assessment Team) (1993). *Forest Ecosystem Management: An Ecological, Economic, and Social Assessment.* Report of the Forest Ecosystem Management Assessment Team. Portland and Washington, DC: US Forest Service, National Marine Fisheries Service, Bureau of Land Management, Fish and Wildlife Service, National Parks Service, and Environmental Protection Agency.

Fleishman, E. D., Murphy, D. & Brussard, P. F. (2000). A new method for the selection of umbrella species for conservation planning. *Ecological Applications*, **10**, 569–79.

Gende, S. M., Edwards, R. T., Willson, M. F. & Wipfli M. S. (2002). Pacific salmon in aquatic and terrestrial ecosystems. *BioScience*, **52**, 917–28.

Giles, N. (2005). *The Nature of Trout*. Dorset: Perca Press.

Groves, C. R. (2003). *Drafting a Conservation Blueprint: A Practitioner's Guide to Regional Planning for Biodiversity*. Washington, DC: Island Press.

Groves, C. R., Jensen, D. B., Valutis, L. L. *et al.* (2002). Planning for biodiversity conservation: putting conservation science into practice. *BioScience*, **52**, 499–512.

Higgins, J. V. (2003). Maintaining the ebbs and flows of the landscape – conservation planning for freshwater ecosystems. In *Drafting a Conservation Blueprint: A Practitioner's Guide to Regional Planning for Biodiversity*, ed. C. R. Groves. Washington, DC: Island Press, pp. 289–318.

Higgins, J. V., Bryer, M., Khoury, M. L. & FitzHugh, T. (2005). A freshwater classification approach for biodiversity conservation planning. *Conservation Biology*, **19**, 432–45.

Holmes, N. T. H., Boon, P. J. & Rowell, T. A. (1999). *Vegetation Communities of British Rivers: A Revised Classification*. Peterborough: Joint Nature Conservation Committee.

Hughes, M., Hornby, D. D., Bennion, H. *et al.* (2004). The development of a GIS-based inventory of standing waters in Great Britain together with a risk-based prioritization protocol. *Water, Air and Soil Pollution: Focus*, **4**, 73–84.

Hunter, M. L. (2001). *Fundamentals of Conservation Biology*, 2nd edn. Malden, MA: Blackwell Science, Inc.

Illies, J. (1978). *Limnofauna Europaea*. Stuttgart: G. Fischer Verlag.

IUCN Species Survival Commission (2000). *IUCN Red List Categories and Criteria*. Version 3.1. As Approved by the 51st Meeting of the IUCN Council, Gland, Switzerland. Gland, Switzerland: The World Conservation Union.

Jelks, H. L., Walsh, S. J., Burkhead, N. M. *et al.* (2008). Conservation status of imperiled North American freshwater and diadromous fishes. *Fisheries*, **33**, 372–407.

Karr, J. R. (1981). Assessment of biotic integrity using fish communities. *Fisheries*, **6**, 21–7.

Kernan, M., Hughes, M., Hornby, D. *et al.* (2004). The use of a GIS-based inventory to provide a regional assessment of standing waters in Great Britain sensitive to acidification from atmospheric deposition. *Water, Air and Soil Pollution: Focus*, **4**, 97–112.

Knight, A. T., Cowling, R. M. & Campbell, B. M. (2006). An operational model for implementing conservation action. *Conservation Biology*, **20**, 408–19.

Lerner, J., Cochran, B. & Michalak, J. (2006). *Conservation Across the Landscape: A Review of the State Wildlife Action Plans*. Washington, DC: Defenders of Wildlife (www.defenders.org/statewildlifeplans/report.pdf).

Maitland, P. S. (1985). Criteria for the selection of important sites for freshwater fish in the British Isles. *Biological Conservation*, **31**, 335–53.

Master, L. L., Flack, S. R. & Stein, B. A. (1998). *Rivers of Life: Critical Watersheds for Protecting Freshwater Biodiversity*. Arlington, VA: The Nature Conservancy (www.natureserve.org/publications/riversOflife.jsp).

Mawle, G. W. & Milner, N. J. (2003). The return of salmon to cleaner rivers – England and Wales. In *Salmon on the Edge*, ed. D. Mills. Oxford: Blackwell Science, pp. 186–99.

Mayden, R. L. & Wood, R. M. (1995). Systematics, species concepts and the evolutionary significant unit in biodiversity and conservation biology. In *Evolution and the Aquatic Ecosystem: Defining Unique Units in Population Conservation*, ed. J. L. Nielsen, American Fisheries Society Symposium 17. Bethesda, MD: American Fisheries Society, pp. 58–113.

Millennium Ecosystem Assessment (2005). *Ecosystems and Human Well-Being : Current State and Trends: Findings of the Condition and Trends Working Group*, eds. R. Hassan, R. Scholes & N. Ash. Washington, DC: Island Press.

Moyle, P. B. & Ellison, J. (1991). A conservation-oriented classification system for California's inland waters. *California Fish and Game*, **77**, 161–80.

Moyle, P. B. & Yoshiyama, R. M. (1994). Protection of aquatic biodiversity in California: a five-tiered approach. *Fisheries*, **19**, 6–18.

Nature Conservancy Council (1984). *Nature Conservation in Great Britain*, Peterborough: Nature Conservancy Council.

Nature Conservancy Council (1989). *Guidelines for Selection of Biological SSSIs*, Peterborough: Nature Conservancy Council.

Nehlsen, W., Williams, J. E. & Lichatowich, J. A. (1991). Pacific salmon at the crossroads: stocks at risk from California, Idaho, Oregon and Washington. *Fisheries*, **16**, 4–21.

Olson, D. M., Dinerstein, E., Wikramanayake *et al.* (2001). Terrestrial ecoregions of the world: a new map of life on Earth. *BioScience*, **51**, 933–8.

O'Sullivan, P. (2005). On the values of lakes. In *The Lakes Handbook. Vol. 2. Lake Restoration and Rehabilitation*, eds. P. E. O'Sullivan & C. S. Reynolds. Oxford: Blackwell Publishing.

Paine, R. T. (1966). Food web complexity and community stability. *American Naturalist*, **100**, 65–75.

Palmer, M. (1992). *A Botanical Classification of Standing Waters in Great Britain and a Method for the Use of Macrophyte Flora in Assessing Changes in Water Quality Incorporating a Reworking of Data.* Research and Survey in Nature Conservation, No. 19. Peterborough: Joint Nature Conservation Committee.

Palmer, M. A. (1999). The application of biogeographical zonation and biodiversity assessment to the conservation of freshwater habitats in Great Britain. *Aquatic Conservation: Marine and Freshwater Ecosystems*, **9**, 179–208.

Palmer, M. A., Bell, S. A. & Butterfield, I. (1992). A botanical classification of standing waters in Britain: applications for conservation and monitoring. *Aquatic Conservation: Marine and Freshwater Ecosystems*, **2**, 125–43.

Pflieger, W. L. (1989). *Aquatic Community Classification System for Missouri*. Aquatic Series No. 19. Jefferson City, MO: Missouri Department of Conservation.

Pollard, P. & Huxham, M. (1998). The European Water Framework Directive: a new era in the management of aquatic ecosystem health? *Aquatic Conservation: Marine and Freshwater Ecosystems*, **8**, 773–92.

Preston, C. D., Pearman, D. A. & Dines, T. D. (2002). *New Atlas of the British and Irish Flora – An Atlas of Vascular Plants of Britain, Ireland, the Isle of Man and the Channel Islands*. Oxford: Oxford University Press.

Ratcliffe, D. A. (ed.) (1977). *A Nature Conservation Review*. Cambridge: Cambridge University Press.

Raven, P. J., Holmes, N. T. H., Dawson, F. H. *et al.* (1998). *River Habitat Quality – the Physical Character of Rivers and Streams in Britain and Isle of Man*. River Habitat Survey Report No. 2. Bristol: Environment Agency.

Richter, B. D., Braun, D. P., Mendelson, M. A. & Master, L. L. (1997). Threats to imperiled freshwater fauna. *Conservation Biology*, **11**, 1081–93.

Rodwell, J. S., Pigott, C. D., Ratcliffe, D. A. *et al.* (1995). *British Plant Communities. Vol. 4. Aquatic Communities, Swamp and Tall-herb Fens*. Cambridge: Cambridge University Press.

Rowan, J. S., Carwardine, J., Duck, R. W. *et al.* (2006). Development of a technique for Lake Habitat Survey (LHS) with applications for the European Union Water Framework Directive. *Aquatic Conservation: Marine and Freshwater Ecosystems*, **16**, 637–57.

Seaber, P. R., Kapinos. F. P. & Knapp, G. L. (1987). *Hydrologic Unit Maps*. Water Supply Paper 2294. US Geological Survey, Denver, Colorado.

Simon, T. P. (ed.) (1998). *Assessing the Sustainability and Biological Integrity of Water Resources Using Fish Communities*. Boca Raton, FL: CRC Press, Inc.

Smith, I. R. & Lyle, A. A. (1979). *Distribution of Freshwaters in Great Britain*. Cambridge: Institute of Terrestrial Ecology.

Smith, R. K., Freeman, P. L, Higgins, J. V. *et al.* (2002). *Priority Areas for Freshwater Conservation Actions: A Biodiversity Assessment of the Southeastern United States*. Arlington, VA: The Nature Conservancy (http://conserveonline.org/docs/2003/08/se_biodiv_assess.pdf).

Smol, J. P. (2008). *Pollution of Lakes and Rivers. A Paleoenvironmental Perspective*. Oxford: Blackwell Publishing.

Sowa, S. P., Anis, G., Morey, M. E. & Diamond, D. D. (2007). A GAP analysis and comprehensive conservation strategy for riverine ecosystems of Missouri. *Ecological Monographs*, **77**, 301–34.

Stein, B., Adams, J., Master, L., Morse, L. & Hammerson, J. (2000). A remarkable array: species diversity in the United States. In *Precious Heritage: The Status of Biodiversity in the United States*, eds. B. A. Stein, L. S. Kutner & J. S. Adams. Oxford, UK: Oxford University Press, pp. 55–92.

Stein, B. & Davis, F. (2000). Discovering life in America: tools and techniques of biodiversity inventory. In *Precious Heritage: The Status of Biodiversity in the United States*, eds. B. A. Stein, L. S. Kutner & J. S. Adams. Oxford, UK: Oxford University Press, pp. 19–53.

Stewart, N. F. & Church, J. M. (1992). *Red Data Books of Britain and Ireland: Stoneworts*. Peterborough: Joint Nature Conservation Committee.

Stewart, A., Pearman, D. A. & Preston, C. D. (1994). *Scarce Plants in Britain*. Peterborough: Joint Nature Conservation Committee.

USFWS (2007). *Aquatic Summaries and Highlights*. A review of Wildlife Action Plans: opportunities to advance fresh water aquatic fish and mollusk species/habitat conservation (www.fws.gov/fisheries/PDFs/Wildlife%20Action%20Plans% 20Aquatic%20Summary.pdf).

Waples, R. S. (1991). Pacific salmon, *Oncorhynchus* spp., and the definition of 'species' under the Endangered Species Act. *Marine Fisheries Review*, **53**, 11–22.

Weitzell, R. E., Khoury, M. L, Gagnon, P. *et al.* (2003). *Conservation Priorities for Freshwater Biodiversity in the Upper Mississippi River Basin.* Arlington, VA: NatureServe and The Nature Conservancy (www.natureserve.org/publications/upperMSriverbasin.jsp).

Wigginton, M. J. (ed.) (1999). *British Red Data Books 1. Vascular Plants*, 3rd edn. Peterborough: Joint Nature Conservation Committee.

World Resources Institute in collaboration with the United Nations Environment Programme, The United Nations Development Programme, and the World Bank (2000). *World Resources 2000–2001: People and Ecosystems – The Fraying Web of Life*. Washington, DC: World Resources Institute.

5 · *Responding to environmental threats within the UK and North America*

CHRISTOPHER A. FRISSELL AND
COLIN W. BEAN

Introduction

The total number of species on the planet is unknown, although an estimate of 15 million species is the most widely accepted figure (IUCN, 2006). Of these, only 1.7–1.8 million species have been fully described by science or are known to exist. Despite accounting for 0.01% of the world's water resources and less than 1% of the Earth's surface, fresh waters support at least 100 000 of these – almost 6% of all described species (Dudgeon *et al.*, 2006). Despite our apparent lack of knowledge of global biodiversity, the *Globally Threatened Species Assessment* (IUCN, 2006) concluded that the number of threatened species is increasing across all taxonomic groups. Under greatest threat are those species that are at the limit of their normal range, or are isolated (e.g. on islands and mountains) and cannot relocate. However, species that have specialized ecological niches reproduce slowly, are reproductively isolated or have restricted gene pools and are also considered to be particularly vulnerable. The latest *Convention on Biological Diversity Report* (Convention on Biological Diversity, 2006) suggests that species' extinction rates are now around 100 times greater than those shown in fossil records.

Several factors have been implicated in the loss of global biodiversity. Human influences, either directly or indirectly, are believed to be the primary reason for the decline or loss of most species. Factors such as habitat destruction and degradation are considered to be the main cause of species decline, although issues such as introduced invasive species (Perrings, 2002; Park, 2004; Genovesi, 2005), unsustainable harvesting, over-hunting, pollution and disease (Smith *et al.*, 2006) also contribute to this loss (Bräutigam & Jenkins, 2001; Perrings, 2002; Park, 2004; IUCN, 2006). Thomas *et al.* (2003), when applying species–area relationships to the current distribution and climatic requirements of 1103 species,

Assessing the Conservation Value of Fresh Waters, ed. Philip J. Boon and Catherine M. Pringle. Published by Cambridge University Press.

concluded that 15–37% of all species in the regions considered in their study could be driven extinct from the climate change that is likely to occur between now and 2050.

High rates of species endangerment and loss indicate particularly acute and accumulating or unrelenting environmental problems afflicting freshwater ecosystems. The most comprehensive study of global freshwater diversity was co-ordinated by the World Conservation Monitoring Centre (WCMC), the United Nations Environment Programme (UNEP) and the IUCN in 1998 (Groombridge & Jenkins, 1998). This study concluded that at high taxonomic levels the diversity of aquatic organisms and overall species richness are considerably narrower than that found in terrestrial or marine habitats, but, in relation to habitat extent, species richness is extremely high within many freshwater groups. Global freshwater biodiversity is now considered to be declining at a rate which is faster than that observed in even the most disturbed terrestrial ecosystems (Revenga *et al.*, 2005). From an ecological perspective, an understanding of the relationship between biodiversity and the functioning of community and population processes in aquatic ecosystems is the key to halting this decline (Jacobsen, 2004).

Threats to aquatic biodiversity in Europe

In 1998, the European Commission published its first *European Biodiversity Strategy* (European Commission, 1998). This strategy aims to reverse present trends in biodiversity loss and place species and ecosystems, including agro-ecosystems, at a satisfactory conservation status, both within and beyond the territory of the European Union by 2010. As a result of this new framework biodiversity objectives are, for example, integrated into all new strategies (such as the European Sustainable Development Strategy), as well as an expanding range of public sector policies. Most Member States have developed, or are developing, their own national biodiversity strategies as part of the wider European framework.

Perhaps unsurprisingly, Europe's ecosystems have experienced more anthropogenic-mediated fragmentation than those of any other continent (European Commission, 2006). Only 1–3% of Western Europe's forests can be classed as being 'undisturbed by humans' and, since the 1950s, Europe has lost more than 50% of its wetlands and high conservation value farmland. Many of the EU's marine ecosystems are also degraded and evidence of damage can be found in even remote coastal areas. At the species level, 42% of Europe's native mammals, 43% of birds, 45% of butterflies, 30% of amphibians, 45% of reptiles and 52% of freshwater fish are threatened with extinction. Most of Europe's major marine fish stocks are considered to be at levels which are below safe biological limits and approximately 800 plant species in Europe are at risk of global extinction. Biodiversity loss within 'lower' taxa such as invertebrates and microbes is relatively unknown.

Despite the small size of the UK compared with Europe, basic data relating to the distribution and status of many of its indigenous species are lacking. This paucity of historic information has made it difficult to determine the conservation status of many taxonomic groups, and it is only recently that the importance of genetic, or intraspecific, diversity has been fully appreciated (for a review, see Frankham, 2005).

Threats to aquatic biodiversity in North America

North America is no exception to the global pattern of high incidence of endangerment in freshwater biota (Allan & Flecker, 1993; Stuart *et al.*, 2004). Amphibians, fish (Frissell, 1993; Warren & Burr, 1994) and mussels (Williams *et al.*, 1993) in the USA show a pattern of extinction and threat exceeding that seen in birds and mammals, and in many taxa of tropical forests. Thirty species of frogs, toads and salamanders are protected as threatened or endangered or listed as candidates for protection under the US Endangered Species Act, with many other taxa in decline or depressed from historical abundance (Lanoo, 2005). The proportional incidence of extinction and endangerment among freshwater fish in the USA is similarly high (Frissell, 1993), with hundreds of fish species, subspecies and populations or stocks recognized by biologists as at risk of extinction (Warren & Burr, 1994).

The primary syntheses of conservation status of US freshwater biota (Williams *et al.*, 1989; Warren & Burr, 1994) list a common and recurring set of threats to species in the past, in the future or in both. These include commercial, sport and subsistence fishing, dams and flow diversions and related flow or stage alteration, land-use change in catchments and riparian corridors, extraction or alteration of stream flows by dams, diversions and groundwater withdrawal, channelization, revetment, drainage, and infilling for floodwater management and floodplain or shoreline development, displacement of native species by invasive non-native competitors, predators and pathogens, and hatchery or aquaculture operations that alter the genetic makeup and adaptive capacity of organisms. Land use or 'habitat loss' are often cited, but are grossly generalized categories that encompass a broad array of human actions and physical and chemical causal chains. Essentially all categories of human development are implicated: transportation systems, urbanization, industrialization, cropland agriculture, grazing, forestry, mining, recreational development, water abstraction for agricultural, domestic and industrial uses and others. As patterns of human development change, the expression of threats shifts. Fresh waters over vast land areas of western North America are principally threatened by grazing, water extraction and logging; in other areas of eastern North America and the centres of urban growth in the West, recreational development, residential and rural residential development have replaced grazing, logging and mining as the principal future threats.

While the USA is a developed country, it still retains some relatively large natural areas that today provide effective 'islands' of high-quality freshwater habitat for many freshwater species (Sedell *et al.*, 1990; Moyle & Yoshiyama, 1994; Frissell & Bayles, 1996; Trombulak & Frissell, 2000). In these areas dominated by natural conditions and processes, the disturbances imposed by civilized man have not seen their full expression on the landscape and its aquatic habitats. By contrast, European ecosystems have seen repeated swaths of human transformation of landscapes over many millennia (Bravard *et al.*, 1986; Brown, 1997). Deforestation, pulses of mining-related sediment and metals, massive channel works and other actions completed hundreds or thousands of years ago express legacies that still strongly shape these ecosystems today (Amoros *et al.*, 1987). Quite likely, many sensitive native freshwater species in Europe passed into extinction during these episodes before they were described by science (Zwick, 1992).

Invasive species, interacting with native species via predation, competitive displacement, ecosystem disruption, transfer of disease or parasites, or interbreeding and hybridization, have been identified as a primary cause of extinction in over half of all endangered species in the USA (Lawler *et al.*, 2002). Freshwater ecosystems are particularly vulnerable to invasion and ecosystem perturbation and reorganization is a consequence of such invasions. Oligotrophic lakes (Spencer *et al.*, 1991) and small streams and rivers with groundwater-dominated, seasonally stable hydrographs (Cavallo, 1997) appear to be highly vulnerable to invasion even when physically undisturbed. Human alteration of natural flow fluctuations by flow diversions or dams, as well as alteration of sediment and flow conditions by disturbance of watershed soil and cover, appear to facilitate or hasten invasion.

The ecology of threats: their role in ecosystem dynamics

Four major conceptual principles have emerged in freshwater conservation science concerning threats to ecosystems and biota, and how they might be managed. Although these have arisen (and are discussed here) in the North American context, these principles are equally applicable to the UK. First, **multiplicity**: *threats do not commonly act as single events, conditions or categories.* The current condition of ecosystems is shaped by their cumulative response to repeated or sustained events against the legacy of prior events, and these usually represent multiple classes of impact (Harding *et al.*, 1998). Teasing out the specific causes of specific conditions mechanistically is usually complicated, frustrating and perhaps sometimes technically impossible. Second, **convergence**: *different kinds of impacts or threats tend to exert common responses in ecosystems* (Schindler, 1987). Hence, even though threats may be diverse and complex in the way they interact, there are often predictable biological outcomes that can be used, like syndromes in medicine, to gauge ecosystem condition and vulnerability. Third, **irreversibility**: the *complexity and systematic effect of ecological changes in fresh waters are often so great that ecosystem responses are virtually or practically irreversible.* Biological invasions, species extinctions, contamination by persistent toxins and massive alterations of catchment land cover and erosion regime are examples. Fourth, **spatial dynamics**: the above ecological outcomes are commonly expressed on the landscape in characteristic geographical patterns, which affords both opportunity and need for the management of threats in a spatially explicit way. In other words, *landscapes and fresh waters can be classified according to the manifestation of threats and their effect on potential future performance of ecosystems*, and priorities in conservation management assigned accordingly. With rare exceptions, only in the past decade or two have aquatic conservation efforts in the USA assumed regional scope in the assessment of biotic integrity and the assigning of conservation priority accordingly.

Current regulatory structures in the USA are largely based on the tacit assumption that threats to fresh waters are easily identified and isolated, decoupled from other elements such as headwater stream channels (Lowe & Likens, 2005) and can be treated by simple interventions or adaptations in such a way that recovery is anticipated with every reduction in presumed impact (Frissell & Bayles, 1996; Espinosa *et al.*, 1997). However, while it is increasingly feasible to describe and measure biological declines in freshwater ecosystems (Schindler, 1987; Karr and Chu, 2000; Karr & Yoder, 2004), reversing those declines through management response is anything but straightforward.

That multiple factors have plagued past efforts at conservation of freshwater resources has not entirely escaped notice by resource managers. For example, an unnamed fishery biologist with the Oregon Department of Fish and Wildlife filed a report in 1948 concerning the status of salmon and steelhead runs in Oregon's Umpqua River. The report hinted at a prolonged and dramatic history of political strife and bloodletting that finally resulted in the closure of a large commercial in-river gillnet fishery, which the managers had come to view as the principal threat to Umpqua salmon stocks. Fishers, in response, pointed to extensive and unregulated logging, land clearance for grazing and crops, and stream diversion, channelization and wetland loss that were continuing in the basin during this time, but the managers prevailed in shutting down the fishery. Salmon returns continued to decline steadily, surprising biologists, as if the fishery had never existed and had not been closed. 'The cessation of the fishery', the report reluctantly concluded, 'exposes the importance of other factors in the decline of the runs'.

Prerequisites for countering threats to aquatic ecosystem conservation

The ability to assess conservation value

This section briefly introduces the role of evaluation frameworks in addressing threats to freshwater ecosystems. More specific accounts of evaluation techniques for rivers and lakes are provided in Chapters 7 and 8.

Evaluating the importance of wildlife and their supporting habitats is a complex issue, made more difficult when supporting data are limited (Abell *et al.*, 2002; Darwall & Vié, 2005). In contrast to terrestrial systems, freshwater habitats and species have received relatively little conservation attention and, as a result, are poorly protected (see Chapter 4). Much of this is due to the fact that basic data relating to distribution and spatial abundance are lacking for all but a few well-studied species. Freshwater habitats are, in general, better understood, but gaps still remain.

In North America, by contrast, the ecology of many threatened species is well understood because of a long history of management data and of academic and agency research. Scientific principles of freshwater conservation are published and well recognized (Stanford & Ward, 1992; Doppelt *et al.*, 1993) and evaluation is largely formalized under the auspices of legislation such as CWA, NEPA, ESA (see Chapter 2) and their counterparts.

From a global perspective, the IUCN Red List has historically allocated individual species to conservation categories based on a complex set of criteria. These may include features such as population size, distribution, rate of decline, extent of fragmentation into sub-populations and generation times (Gardenfors *et al.*, 1999). Although some species are well studied throughout their range and can be accurately ascribed a 'conservation category', ranging from 'extinct', or 'extinct in the wild' to 'least concern', a significant number of freshwater species are still listed as being 'data deficient' in global terms. The IUCN takes, as one of its main data sources, the Red Data lists of various countries across the globe. For countries that have a coordinated biodiversity surveillance strategy in place, such datasets are invaluable. However, many countries do not have such measures in place at all, and others may have information on selected taxa only. Within the UK, for example, Red Data lists exist for a variety of terrestrial invertebrates and vascular plants, but no such list is available for freshwater fish.

A legislative framework

Chapter 2 summarizes the main areas of national and international legislation relevant to freshwater conservation and further information relating to the legislative protection afforded to rivers and lakes is provided in Chapters 7 and 8. This section emphasizes those aspects of the UK and the US legislative frameworks which are especially relevant to responding to environmental threats.

Despite having in place a strategy of protecting sites of scientific or natural heritage importance, it is clear that in the UK significant habitat loss and damage to individual species continue to take place. Much of this has, historically, been due to pressures within the agricultural and forestry sector, but also from built development. Latterly, the drive towards renewable energy sources has, ironically, led to greater pressure on some elements of the natural heritage. This forced the UK government to review its nature conservation policies and led to calls to extend the levels of protection afforded to these sites and to bring UK legislation in line with relevant EC nature conservation legislation. Over 80% of the UK's environmental legislation currently has its origins within European Directives and it is clear that the UK's conservation policy will continue to be led by wider, European-level, objectives.

The UK has a responsibility to ensure the conservation of habitats and species in both a national and international context. Within the UK, the

Wildlife and Countryside Act 1981 (as amended) was, until recently, the primary legislative tool for protecting the natural heritage. This legislation provided the basis for protecting a number of named plant and animal species as well as the establishment of SSSIs. At present, there are 6569 Sites of Special Scientific Interest (SSSIs) in Great Britain, with another 225 Areas of Special Scientific Interest (ASSIs) established in Northern Ireland. Between them, these designations protect approximately 2.4 million hectares of terrestrial and aquatic habitat. Fewer sites (210) are also protected through the National Nature Reserve (NNR) series, and a network of smaller, non-statutory, Local Nature Reserves (LNRs).

More recently, The Nature Conservation (Scotland) Act 2004 and, to a lesser extent, the Countryside and Rights of Way Act 2000 (in England and Wales) have strengthened these provisions. In the case of the Nature Conservation (Scotland) Act 2004 this legislation has, for the first time, placed a statutory duty on all public bodies to further the conservation of biodiversity. This means that public bodies must examine not only how they run their operations and incorporate actions to conserve biodiversity but also how their functions can help to deliver biodiversity conservation objectives. This has clear implications both for the local authorities and for other public bodies that may have a role in the planning process.

In the USA, the basic legislation that provides the foundation of freshwater conservation actions reflects a trend of the federal government imposing more restrictive and proscriptive regulatory frameworks on states, local governments and private developers as public concern increased about deterioration of water, air and land. The traditional legal and regulatory mechanisms in the states were focused on the expedient allocation of resources like fish and water for their extractive use. The increased federal regulatory burden was accompanied by substantial federal funding and expert technical support to help states, businesses and others implement conservation programmes. Some US laws simply withdraw specific tracts of land from unfettered allocation, in effect creating some level of natural reserve. Examples include the early establishment of a national forest reserve and a national park system, and subsequent laws established wilderness areas, parks or wildlife refuges. However, these areas were very seldom selected or consciously designed and managed to overtly protect freshwater biological resources. Because of factors such as fish stocking, man-made migration barriers and manipulation of waterways for specific use by waterfowl or other game species, their function as reserves for aquatic biota is to some degree compromised (Williams, 1993).

Other federal laws create performance benchmarks or planning processes intended to ensure that a fuller spectrum of environmental values is recognized as natural resources are developed and managed; some, such as the National Forest Management Act of 1976, mention freshwater values, but only the Clean Water Act (CWA) of 1977 specifically focuses on aquatic ecosystems. However, it can be argued that even the CWA is primarily a means of regulating practices that threaten surface waters with direct harm. While it established 'goals', it is not generally interpreted to confer a sweeping mandate or unambiguous, explicit authority to ensure that aquatic ecosystems are conserved in the biological sense of the word (Doppelt *et al.*, 1993). Simply put, no law (with the possible exception of the ESA in limited circumstances) establishes the primacy of conserving fresh waters over the rights and traditions of development and extractive land use, such as grazing, logging and mining. As a result of freshwater and numerous other species suffering declines under other laws, the Endangered Species Act of 1976 has played an increasingly important role as a conservation 'safety net'. Linkage of the benchmarks or directives in the ESA and other laws to the planning and disclosure provisions of NEPA creates a complex and tenuous legal fabric against which management of catchments and waters is carried out.

Protecting aquatic environments through the planning system

Built development and habitat loss through fragmentation or isolation is regarded as being the biggest threat to biodiversity (Park, 2004), and it is clear that the planning system and development control is vital in protecting freshwater habitats and species.

Planning within Europe and the UK

Environmental law and the protection of natural resources within the UK is not a recent development. In fact, provisions have existed for the protection of some habitats (i.e. forests), certain (principally game) species, and even the prevention of water pollution, for many centuries (Howsam, 2003). However, much of these protective statutes were developed on an ad hoc basis and were not part of a wider programme of protective measures. Nature conservation has, since the National Parks and Access to the Countryside Act 1949, been a core element of domestic planning law within the UK. However, devolved administration and, in particular the persistence of a separate Scottish legal system, means the mechanism by which this is delivered may differ markedly between countries. Generally, however, planning policy at a local and national level is enshrined within a series of Structure Plans and Local Plans that set out a broad strategy, and land-use policies for each of the local authorities within the UK. Structure plans should take account of appropriate published national policy guidance and place particular emphasis on the strength of protection afforded to international and national natural heritage designations. Such plans should also take full account of the implications for the natural heritage in considering possible locations for new strategic development. It should also seek to identify strategic opportunities both for enhancing the natural heritage and deriving social and economic benefits from it.

The UK government and the devolved administrations of Scotland, Wales and Northern Ireland recognize that the application of environmental appraisal techniques in the planning process can make an important contribution to the protection and enhancement of freshwater habitats and species. Much of this is achieved through the publication of a series of planning circulars, guidelines and advice notes, which ensure that local planning authorities take account of natural heritage interests in a manner that is consistent both within individual countries and across the UK. More recently, the key driver in the protection of freshwater habitats and species is the EC Water Framework Directive (WFD; see Chapter 2). The WFD incorporates the provisions of existing conservation legislation and the Habitats and Birds Directives. The new regulatory regimes required by the WFD establish measures to regulate activities for the purposes of protecting the water environment, and will facilitate the achievement of environmental objectives set out in river basin management plans.

In addition to the WFD, the Strategic Environmental Assessment (SEA) Directive is designed to make the planning process more transparent by extending opportunities for participation in public policy decision-making. The SEA process requires all new 'Plans, Projects and Strategies' (PPS) to be systematically assessed and requires the environmental effects of any new ones to be monitored. It also places an obligation on policy makers to seek the views of statutory conservation agencies and environmental regulators. By integrating environmental and public considerations into the policy process, government and local planning authorities may now be able to address some of the more generic issues that affect freshwater habitats and species on a wider scale.

While in some circumstances it will be necessary to refuse planning permission for a given development on natural heritage grounds, authorities should always consider whether environmental concerns could be adequately addressed by modifying the development proposal or attaching appropriate planning conditions. To inform this process, planning authorities may, for a number of development types, require the provision of an Environmental Impact Assessment (EIA). The EIA concept, currently in place within the UK and other parts of Europe, has its roots in North America. There, the 1970 US National Environmental Policy Act (NEPA), was the first formal national legislation to require the delivery of EIAs and this, in broad terms, became the model for other countries around the world. Within Europe, EIA theory and practice was not formally introduced to Member States until 1985, when Directive 85/337/EEC set out a broad framework for EIA production. The development of formal legislation relating to the EIA process has been similarly slow to develop within the UK and has only been in place since 1988. The impetus for the development of EIA-specific domestic legislation was the need to transpose EC Directive 85/337/EEC into UK law. Glasson *et al.* (1999) suggest that, without pressure from the European Commission, formal legislation would never have been developed and that the UK would have continued to rely on the ad hoc arrangement that existed within the Town and Country Planning legislation.

To say that EIAs were not used within the UK prior to 1988, however, would be untrue. Since 1947, the UK relied primarily on the provisions of its statutory land-use planning system and EIAs were prepared either voluntarily or at the specific request of local planning authorities. These failed to follow any specific format or statutory guidelines. By the 1960s the relationship between statutory planning controls and the impact of large development came under considerable scrutiny, although at that time more attention was focused on socio-economic, rather than environmental, impacts. In common with the development of EIA procedures in North America, the North Sea oil boom provided the biggest push towards the development of environmental assessment within the UK.

The European EIA Directives are relatively straightforward in terms of their structure and guidance. Put simply, development of a type listed in Annex I of the Directive (e.g. refineries or power stations) *always* require the production of an EIA. Developments listed in Annex II require EIA 'if it is likely to have significant effects' on the environment by virtue of factors such as their size, nature or location. Within the UK special considerations apply to SSSIs, especially those that are also used to underpin international conservation sites. In practice, the likely environmental effects of Annex II development will often be such as to require EIA if it is to be located in or close to sites classified as Special Protection Areas (SPAs) or Special Areas of Conservation (SACs) under the EC Habitats and Birds Directives, or Ramsar sites (Hoskin & Tyldesley, 2006).

Where an EIA shows that a development or activity may have a negative impact on habitats or species of conservation importance, local planning authorities may consider whether planning conditions or legal agreements may mitigate these impacts sufficiently to allow the development to proceed. It remains, however, the task of the local planning authority to judge each planning application on its merits within the context of their own Structural and Local Development Plan. This should take account of all material considerations, including the environmental impacts, of any development proposal.

Case study: EIA and hydropower development in north-west Scotland

The proposed Sheildaig and Slattadale hydro scheme involved the impoundment of three lochs within the Wester Ross National Scenic Area in the north west of Scotland. Some locations within the area are currently utilized by other hydro developers. In addition to its scenic value, the affected area is important for a range of other conservation interests. These include the River Kerry SSSI and SAC, important for its population of freshwater pearl mussel *Margaritifera margaritifera* and Atlantic salmon *Salmo salar*; the Loch Maree SSSI/SPA/ Ramsar (wetland of national importance), important for its water and woodland habitat, its bird and invertebrate species and its geology; and the Loch Maree Complex SAC, important for its variety of plant habitats and otters.

Scottish Natural Heritage (SNH) and others objected to this proposal on the grounds that the development may have a significant negative impact on species protected under Annex II of the EC Habitats Directive (freshwater pearl mussel *Margaritifera margaritifera*), Annex I of the EC Birds Directive (black throated diver *Gavia arctica*) and landscape. Other issues, such as the impact on freshwater fish (e.g. brown trout *Salmo trutta* and Arctic charr *Salvelinus alpinus*), were also cited. The Scottish Environment Protection Agency (SEPA) also objected to the development on the basis that it would have adverse hydrological impacts that would, in turn, affect the aquatic flora and fauna. The scheme was of moderate size (3.55 MW) and was first submitted in 1996. It was subsequently withdrawn during a Public Inquiry in 1997 – before the Inquiry Reporter had made a judgement. In 2002, the developer submitted a revised, and expanded, plan which included the impoundment of a fourth loch.

The Scottish government is responsible for determining applications for consent under the Electricity Act 1989. These are required for electricity generating stations with an installed capacity greater than 50 MW, or greater than 1 MW in respect of hydroelectricity. Schedule 3 of the EIA Regulations outlines which factors ministers should consider in arriving at a determination. These include:

- characteristics of the development (size, use of natural resources, waste pollution);
- location of the development (existing land-use, regenerative capacity of natural resources and the capacity of the natural resources to absorb the development);
- characteristics of the potential impact (extent, magnitude and complexity of the impact, probability of the impact, duration and reversibility of the impact).

Following receipt of the EIA for this development, the Scottish ministers decided that, in their view, a Public Local Inquiry (PLI) was not required because they already had enough information on which to base their determination. This application was refused by Scottish ministers in 2004 on the grounds that the applicant was given an opportunity to respond to Ministerial concerns earlier in the process and that their final response did not alleviate these concerns. These concerns centred mainly on the potential for the development to have an adverse impact on species of international conservation importance. A decision to refuse the application was therefore taken.

One of the key areas where EIA regulations must focus more attention is the issue of cumulative impact (Piper, 2002). This case study demonstrates that even moderate-sized developments can have a significant negative impact on sites or species if they are considered along with existing structures. The European EIA process is constantly under review and future amendments will reflect new experiences and case law as it occurs within each of the Member States (Commission of European Communities, 2003).

Planning within North America

The principal policy framework for environmental planning and assessment in the USA is the NEPA of 1970. Projects or programmes authorized or financed by the federal government that have sweeping scope and possible extensive, costly or long-term effects, or that represent a sharp departure from past management activities, warrant an intensive analysis known as the Environmental Impact Statement (EIS), whereas projects of more limited scope, spatial scale and duration are subject to a less intensive Environmental Assessment. Hence NEPA is a planning framework that is driven by the mitigation of potential environmental threats; it is not in the broader sense a strategic planning framework. NEPA itself does not mandate conservation or restoration outcomes. NEPA interfaces with other federal laws that govern and establish standards for some specific elements of land and water planning and management, including the CWA, the National Forest Management Act, the Wild and Scenic Rivers Act and the US Endangered Species Act (see Chapter 2; Doppelt *et al.*, 1993).

As a planning framework, NEPA mandates the 'consideration' of best available scientific information, explicit delineation of the human and natural environment affected by a given project, consideration of reasonable alternatives to any proposed action, and full disclosure of the reasoned basis for decisions about which action to undertake. NEPA also mandates some explicit kinds of analysis that are germane to freshwater impacts, but are seldom implemented effectively (Ziemer *et al.*, 1991), including consideration of the foreseeable cumulative and indirect impacts of the proposed action in combination with other past or anticipated actions or events in the affected ecosystems. US government agencies can be sued by citizens and organizations, and projects stopped or delayed, when NEPA planning requirements are not faithfully executed. Some states have adopted similar laws requiring analysis and disclosure of foreseeable impacts of non-federal development.

Case study: EIA and hydropower development in Pacific North America

The Federal Energy Regulatory Commission (FERC) prepares EISs for hydropower dams as part of their re-licensing process in order to fulfil NEPA requirements. A recent example from the Klamath Hydroelectric Project on the Klamath River in the North Pacific state of Oregon illustrates how FERC must decide when considering re-licensing and what conditions to place on any licence that is issued. In deciding whether to authorize the continued operations of the hydropower project, FERC must determine whether the project will be best adapted to a comprehensive plan for improving or developing a waterway. In addition to power and developmental purposes (e.g. flood control, irrigation and water supply), FERC must give equal consideration to the purposes of energy conservation; the protection and enhancement of fish and wildlife (including related spawning grounds and habitat); the protection of recreational opportunities; and the preservation of other aspects of environmental qualities.

The Klamath Hydroelectric Project, owned and operated by PacifiCorp, consists of eight 'developments' or components under re-licensing. Three of these developments were being proposed for decommissioning or removal. Some of these proposals are based on the costs of structural changes required for continued operations in order to protect federally listed fish species. PacifiCorp proposed to operate the five remaining developments similar to past operations with a set of 41 environmental measures, which include specified minimum flows, installations of fish passages, placing gravel for enhancing fish spawning habitat and implementation of habitat management plans, among others.

After review and recommendations of the proposal from resource agencies, Indian tribes and other interested parties, NRCS proposed what were termed 'Staff Alternatives'. These alternatives incorporated most of the PacifiCorp proposed environmental measures with some modifications to them, and an additional 25 environmental measures. These measures included descriptions and details for implementing an integrated fish passage and disease management programme, different minimum flow rates and implementation of an adaptive sediment augmentation programme based on habitat mapping.

Two other sets of alternatives were proposed in addition to those of PacifiCorp and the Staff Alternatives: Staff Alternatives with Mandatory Conditions (alternatives developed as a result of a complex set of multiple federal agency and PacifiCorp filings and responses to decisions on alternatives which would have resulted either in modification or elimination of several Staff Alternatives), and no actions (maintaining the same environmental measures as in the existing licence. Based on a review of all proposals and alternatives, environmental benefits and costs, they decided to issue a new licence they deemed to be consistent with the environmental measures specified in the Staff Alternatives. While the results remain controversial, the rationale and information considered by FERC are documented and available for public and scientific scrutiny.

Protecting the resource – challenging development

The UK experience

If having gone through this process a development is considered to have a negative impact on the environment, and planning permission is refused, developers may have the opportunity to appeal against this decision. In many cases, developers may either abandon their application or seek to present their case at a PLI. The bodies that oversee the PLI process differ between countries within the UK and cases can be heard by the Planning Inspectorate (England and Wales), the Scottish Executive Inquiry Reporters Unit and the Planning Appeals Commission (Northern Ireland). Within Northern Ireland, inquiries relating to water drainage and fisheries are heard by the Water Appeals Commission. There are no equivalent bodies within Scotland, England or Wales. Unfortunately, the mechanism currently used by appeal bodies to collate and report planning appeal data makes it impossible to disaggregate those which involve disputes over development decisions affecting freshwater habitats or species.

Case study: small-scale housing development

When there is a conflict between a development proposal and natural heritage interests of a high conservation value, PLIs tend to attract much public and media interest. This case study is an example of a PLI which involved the construction of two relatively modest dwellings in an area where it would have had a damaging impact on a wide range of freshwater SSSI, SAC and SPA features. It also demonstrates that even relatively small developments may have to go through the PLI process if their impact on sites designated for their important habitats and species is considered to be significant.

The Mawcarse development involved the construction of two dwelling houses on the shores of Loch Leven. The Loch Leven SPA is designated for its internationally important over-wintering populations of swans, geese and ducks, which are migratory or listed on Annex I of the EC Birds Directive. Whooper swan *Cygnus cygnus*, pink-footed goose *Anser brachyrhynchus* and shoveler *Anas clypeata* all overwinter at the loch. In addition to its bird interests, Loch Leven is also the largest naturally eutrophic lake in Britain. It is a relatively shallow loch, surrounded by farmland, with a diverse aquatic flora and shoreline vegetation. Its wide range of biological and habitat interests is reflected in the number of conservation designations conferred upon the site (SSSI, NNR, Ramsar, SPA). It is also a world-renowned brown trout *S. trutta* fishery and this 'unique' genetic strain of trout has been translocated to various locations around the globe since the turn of the twentieth century.

Water quality within the loch had declined considerably in the recent past, with large influxes of phosphorus, both from point and diffuse sources, causing algal blooms. Methods of reducing phosphorus inputs into the loch were identified within a catchment management plan in 1995 and restrictive policies were incorporated into the Local Plan. In the few years preceding the planning application at Mawcarse, this had proved to be a successful strategy and water quality within the loch improved. The developer originally proposed to use a sewage system that was already in place for other dwellings in the area. The proposal was changed to include a septic tank arrangement that would have discharged water to the ground or local watercourses. This would have contributed to the diffuse pollution entering the loch, and the potential phosphorus discharge from two new houses was estimated at 8100 mg day^{-1}.

SNH considered that this proposal *was* likely to have a significant effect upon the conservation interests of Loch Leven SPA and that an Appropriate Assessment would be required. The developer then submitted a mitigation scheme which was considered at the Public Inquiry which proposed installation of a new efficient treatment plant for the new dwellings. This would significantly reduce the phosphorus discharge from the existing dwelling to a level below that currently discharged by neighbouring structures. SNH confirmed that the development proposal would not now have an adverse effect on the integrity of Loch Leven SPA.

The Inquiry Reporter commented that 'Although it was difficult to conclude that the proposal alone would be likely to result in an adverse effect on the loch's integrity, if it was approved without mitigation, an important precedent would be set, which would nullify the aims of the catchment and Local Plans'. The Reporter also made reference to the Waddenzee judgement to give weight to the conclusions drawn, stating that 'A recent European Court of Justice decision relating to a case in the Netherlands (Landelijke Vereniging tot Behoud Van de Waddenzee, Nederlandse v Vereniging tot Bescherming von Vogels v Straatssecretaris Van Landbouw, Natuurbeheer en Visserij (C-127/02: [2005] Env. LR14 [ECJ]) confirmed that where a proposal not directly connected with or necessary to site management was likely to undermine a site's conservation objectives, it would have a significant effect'.

Public Local Inquiries are generally the last recourse for developers to get a contentious planning application through the planning process. If the Inquiry Reporter upholds the case of the local planning authority or relevant ministers, they may seek further consideration at a higher court of appeal. In reality, this course of action is rarely undertaken. Regardless of the outcome, PLIs benefit those bodies charged with protecting aquatic environments simply by providing new examples of case law. This, in turn, clarifies the legal interpretation of domestic and European conservation legislation and provides a firm basis for future protection of vulnerable habitats and species.

The North American experience

In the USA, implementation of conservation statutes for the past quarter-century has been driven by citizen legal appeal of government decisions that courts deemed inadequately reasoned to protect natural resource and biological values. The US Endangered Species Act, in particular, because its requirements for protection and penalties for harm are potentially so strict, has played a prominent role in forcing government to recognize and try to remediate or reduce environmental threats to waters and their biota. ESA listing of numerous formerly wide-ranging freshwater taxa has prompted calls for widespread reform of management of both government-controlled and privately held land and water developments. Laws to protect water quality, restrict fisheries and reform management of federal forest lands, if they met their explicit intentions, should be expected to pre-empt the need for endangered species management. However, whether through shortcomings of those policies, or failure to implement them fully, they have possibly slowed but not significantly reversed the continuing decline of freshwater species in the USA (Fausch *et al.*, 2002; Frissell, 1993; Warren & Burr, 1994).

Case study: implementing protective legislation for the bull trout

The case of the bull trout (*Salvelinus confluentus*) serves as an example of how lengthy this give-and-take process can be when most conservation decisions are extracted by court decision from a reluctant government. To win an ESA case against a federal agency in court, citizens have to demonstrate that the agency has ignored, shirked or erred in executing its duty to protect species and their habitats; this often involves demonstrating misuse of, inattention to, or occasionally, indefensible invention of scientific information.

Once native to a broad region extending from northern California north to British Columbia, Canada and inland to the Rocky Mountain Continental Divide, this species requires clear, cold waters and stable spawning and early rearing habitats to complete its early life history. Migratory adults are dependent on access to rivers and lakes free of obstructions or severe alterations of physical habitat or trophic structure in order to move from foraging to spawning habitat. Bull trout have declined until they currently occupy less than half of the estimated native range, with only 6% of the present range occupied by populations considered demographically secure ('strong') (Rieman *et al.*, 1997).

A petition to list the bull trout under the US Endangered Species Act was first submitted to the US Fish and Wildlife Service in 1992, and it was finally listed as a threatened species eight years later. This listing did not occur until six sequential legal proceedings either forced the agency to act after it exceeded legally mandated time limits on decisions, or reversed agency decisions were deemed illegal. Litigation continues at the time of writing over whether the US Fish and Wildlife Service has met its legal obligation to designate critical habitat for the predominant portion of the species' current distribution, as well as additional habitat that is necessary for its ultimate recovery. Moreover, 14 years after the species was petitioned for listing, the mandated federal recovery plan for the bull trout is not in place.

Despite this foot-dragging from high-level federal officials, state and federal field biologists report that the protected status of bull trout has resulted in modification of numerous projects (roads, dam operations, timber sales, irrigation diversions and others) in ways that significantly reduced their impact on streams and lakes. Biologists and other experts report that public awareness and the day-to-day management affecting bull trout habitats has improved markedly as a direct result of their federal protection (Chris Clancy, pers. comm., Montana Department of Fish, Wildlife & Parks, Hamilton, MT).

The move to collaborative processes

The delivery of environmental monitoring and management objectives within the UK has, for many years, been largely dependent not only on the activities of government departments and agencies, but also on a wider network of non-government organizations (NGOs) and local interest groups. Over the course of the last 20 years, many of these groups have become organized into a larger network of bodies known as 'LINKs'. These LINK bodies, of which there is one in each of the four countries of the UK, represent a broad spectrum of environmental interests and promote communication between the voluntary sector and government. They represent a powerful lobbying force and can directly influence the development of environmental policy and legislation. In doing this, they can, strategically, reduce the risk of threats to habitats and species by ensuring that appropriate measures are in place to counter these should they arise.

NGOs, and LINK bodies in particular, have been extremely important in the protection of aquatic habitats and species in recent years. For example, a review of the aquatic elements of the UK Biodiversity Action Plan (BAP) was largely dependent on the provision of data which had been collected by NGOs and stored in the National Biodiversity Network (NBN) database. The active involvement of NGO and LINK bodies in specialist groups was also instrumental in the delivery of new priority lists for the protection of aquatic species and habitats. Such is the level of organization within NGOs and LINK groups that they are now routinely asked to contribute to the development of catchment management plans and, more recently, river basin management plans under the WFD.

Whilst many of the constituent bodies within each LINK are relatively small, some, such as the RSPB and WWF-UK, are disproportionately well resourced. In fact, the combined annual income of the RSPB and WWF-UK amounts to a figure well in excess of £100 million, exceeding that of some government agencies and non-governmental public bodies (such as SNH). The NGOs within the UK do not restrict their activities to domestic issues, but have formed constructive links with similar bodies in mainland Europe; by doing so, they are actively involved in trying to influence policy and legislation at a European level. By liasing with NGOs based in Brussels, such as the European Environmental Bureau (EEB) and the European Bureau for Conservation and Development (EBCD), UK NGOs can gain direct access to the European Commission and other international organizations such as the IUCN. Through a process of organization and direct involvement, UK NGOs have played, and continue to play, a major role in responding to environmental threat at both a national and an international level.

Recently many government authorities and non-governmental organizations in the USA have pressed for more catchment-focused, so-called 'community-based collaboratives' with the ambition that these arrangements will more effectively address and integrate the full spectrum of threats and management challenges, while avoiding the political costs that come with making hard choices about conservation. Communities and volunteers or citizen activists have always borne much of the burden of monitoring and enforcement of existing environmental protection in the USA, but the new 'collaborative paradigm' extends the expected work of volunteers and citizens to encompass the full span of regulatory activity formerly accomplished by government. In some cases collaborative organizational frameworks have succeeded and achieved measurable and laudable advances on the ground. Many other cases, less commonly discussed, have met with limited success or disappointment.

While appreciating their social virtues, it is important to recognize several critical and recurring limits on the ability of collaboratives to deliver conservation outcomes. First, even when they are sponsored by government, they lack authority to regulate or revise management practices that are widespread and considered routine or fiscally advantageous by large corporations and government agencies (Huntington & Sommarstrom, 2000). Voluntary or consensus-based reforms are commonly ineffective when environmentally harmful 'old ways of business' are culturally engrained or highly profitable. Examples are logging in riparian zones and livestock grazing, which benefit certain businesses, yet cause serious and pervasive harm to aquatic ecosystems; the human costs diffuse to often-distant fishers, and among all other water users. Second, localized collaboratives often suffer from lack of appropriate expert resources and access to relevant knowledge, which limits their ability to diagnose and treat effectively the root causes of harm. Their actions often move forward based on consensus only about threats that are highly obvious and certain; equally acute threats that are controversial or unrecognized, and require specialized or unbiased scientific expertise to diagnose and treat, are left unattended. Third, catchment collaboratives often suffer from loss of continuity of staff and programmes because they lack the dedicated funding that agencies have, and grant funding is unstable. This in turn hinders their ability to develop and retain expert staff. For these reasons, collaborative processes may be intrinsically unsuited to much of the work of freshwater conservation. The more effective catchment collaboratives in the USA appear to recognize these limits and partner or parlay with government agencies to complement their consensus-based work.

Conclusions

Threats to freshwater ecosytems and biota are pervasive, long-standing and often interact or co-occur in complex ways. The extraordinary connectivity of fresh waters to other fresh waters, and to the catchment and surrounding landscape, precludes success of fine-scale and strictly localized protective action. This ecosystem context complicates conservation, and necessitates that effective conservation programmes must be able to act against a wide range of threats that encompass a broad spectrum of human activity and institutions.

In the UK, conservation and the protection of aquatic habitats and species is now being dictated by international (EU) Directives and policies – leading to pan-European monitoring and reporting systems. The Habitats Directive provides a good example of a standardized approach to monitoring, although differences between Member States in how they designate and manage sites within the Natura 2000 site series limit its value. Greater potential lies in the WFD, which has at its heart the development of common sampling and reporting protocols for aquatic environments, in line with a series of pan-European river basin management plans. This may, in part, address previously held concerns about the lack of aquatic habitats and species data which has been a problem for those required to manage these features within the UK. In turn, European initiatives such as the Habitats Directive and the WFD link to global programmes to protect freshwater habitats and species – such as the Convention on Biological Diversity, Agenda 21 and the UN 'Water for Life' programme.

In the USA, a complex fabric of federal environmental laws bears to varying degrees on the conservation of lakes, rivers, streams and wetlands. The CWA addresses freshwater ecological values most directly and broadly, but as implemented, its regulatory effectiveness is limited to the incremental, project-by-project reduction of harm from ongoing polluting activities. Therefore, the reach of the CWA does not extend far enough to ensure ecologically effective conservation management in most instances. State and federal agencies that manage fisheries have also inherited a strong bias for propagating fishing interests, which can sometimes fly in the face of biological conservation and is directly threatened by federal protection of endangered species. The result is that habitat deterioration and species declines have led inexorably to endangered listings for many formerly widespread taxa, forcing more and more waters to be managed under stricter conservation requirements. Vested interests and government agencies are often loth to exercise the terms of the ESA, because of the political cost encumbered by robust protection of freshwater habitats and popular game species, hence they resist extending federal protection to species.

Once species are listed, or in the few cases when political leadership is exerted to take bold steps to head off ecological declines – as in the Northwest Forest Plan, the Kissimmee system in Florida and some other regional projects – regulatory programmes seem to result in increased public awareness, cooperation of threatened interests and avoidance of once-prevalent harm on the ground. These regulatory measures are complemented, or sometimes putatively replaced by, increasingly popular collaborative consensus- and catchment-based initiatives and organizations. In isolation, collaboratives are limited in their ability to attack many entrenched threats, but they can be well suited to others. Where collaboratives and regulatory agencies work together in complementary ways, measurable conservation and restoration benefit to fresh waters and the catchments they depend on can be achieved. Such effective articulation of voluntary and mandatory action is not yet commonly found, however, across the USA.

Acknowledgements

The first author thanks the Pacific Rivers Council for supporting his work on this chapter. We also thank Jonathan Higgins (The Nature Conservancy, Global Conservation Approach Team, Chicago, IL, USA) for his contributions to this chapter.

References

Abell, R., Thieme, M., Dinnerstein, E. & Olson, D. (2002). *A Sourcebook for Conducting Biological Assessments and Developing Biodiversity Visions for Ecoregion Conservation. Volume II: Freshwater Ecoregions.* World Wildlife Fund. Washington, DC: Island Press.

Allan, J. D. & Flecker, A. S. (1993). Biodiversity conservation in running waters. *BioScience*, **43**, 32–43.

Amoros, C., Rostan, J., Pautou, G. & Bravard. J. (1987). The reversible process concept applied to the environmental management of large river systems. *Environmental Management*, **11**, 607–17.

Bräutigam, A. & Jenkins, M. (2001). *The Red Book: The Extinction Crisis.* Published by CEMEX, Mexico, in collaboration with IUCN's Species Survival Commission and Agrupación Sierra Madre.

Bravard, J. P., Amoros, C. & Pautou, G. (1986). Impact of civil engineering works on the successions of communities in a fluvial system: a methodological and predictive approach applied to a section of the Upper Rhône river, France. *Oikos*, **47**, 92–111.

Brown, A. J. (1997). Clearances and clearings: deforestation in Mesolithic/Neolithic Britain. *Oxford Journal of Archaeology*, **16**, 13–146.

Cavallo, B. J. (1997). *Floodplain Habitat Heterogeneity and the Distribution, Abundance and Behavior of Fishes and Amphibians in the Middle Fork Flathead River Basin, Montana.* Missoula: Division of Biological Sciences, University of Montana.

Commission of European Communities (2003). *Five Year Report to the European Parliament and Council on Application and Effectiveness of EIA Directive.* Brussels: CEC.

Convention on Biological Diversity (2006). *Report of the Eighth Meeting of the Parties to the Convention on Biological Diversity*. Eighth meeting, Curitiba, Brazil, 20–31 March 2006. Report UNEP/CBD/COP/8/31, 15 June 2006.

Darwall, W. R. T. & Vié, J. C. (2005). Identifying important sites for conservation of freshwater biodiversity: extending the species-based approach. *Fisheries Management and Ecology*, **12**, 287–93.

Doppelt, B., Scurlock, M., Frissell, C. & Karr, J. (1993). *Entering the Watershed: A New Approach to Save America's River Ecosystems*. Washington, DC: Island Press.

Dudgeon, D., Arthington, A. H., Gessner, M. O. *et al.* (2006). Freshwater biodiversity: importance, threats, status and conservation challenges. *Biological Reviews of the Cambridge Philosophical Society*, **81**, 163–82.

Espinosa, F. A., Rhodes, J. J. & McCullough, D. A. (1997). The failure of existing plans to protect salmon habitat on the Clearwater National Forest in Idaho. *Journal of Environmental Management*, **49**, 205–30.

European Commission (1998). *Communication of the European Commission to the Council and to the Parliament on a European Community Biodiversity Strategy*. COM (98)42, Brussels.

European Commission (2006). *Halting the Loss of Biodiversity by 2010 and Beyond: Sustaining Ecosystem Services for Human Well Being. Impact Assessment*. Council Communication, COM(2006)216, Brussels, 22 May 2006, SEC(2006) 607.

Fausch, K. D., Torgerson, C. E., Baxter, C. V. & Li, H. W. (2002). Landscapes to riverscapes: bridging the gap between research and conservation of stream fishes. *BioScience*, **52**, 483–98.

Frankham, R. (2005). Genetics and extinction. *Biological Conservation*, **126**, 131–40.

Frissell, C. A. (1993). Topology of extinction and endangerment of native fishes in the Pacific Northwest and California, USA. *Conservation Biology*, **7**, 342–54.

Frissell, C. A. & Bayles, D. (1996). Ecosystem management and the conservation of aquatic biodiversity and ecological integrity. *Water Resources Bulletin*, **32**, 229–40.

Gardenfors, U., Rodriguez, J. P., Hilton-Taylor, C. *et al.* (1999). Draft guidelines for the application of IUCN Red List Criteria at national and regional levels. *Species*, **31–2**, 58–70.

Genovesi, P. (2005). Eradications of invasive alien species in Europe: a review. *Biological Invasions*, **7**, 127–33.

Glasson, J., Therivel, R. & Chadwick, A. (1999). *Introduction to Environmental Impact Assessment: Principles and Procedures, Process, Practice and Prospects*, 2nd edn. London: UCL Press.

Groombridge, B. & Jenkins, M. (1998). *Freshwater Biodiversity: A Preliminary Global Assessment*. World Conservation Monitoring Centre Series No. 8. Cambridge: WCMC-World Conservation Press.

Harding, J. S., Benfield, E. F., Bolstad, P. V., Helfman, G. S. & Jones, E. B. D. (1998). Stream biodiversity: the ghost of landuse past. *Proceedings of the National Academy of Sciences of the USA*, **95**, 14843–7.

Hoskin, R. & Tyldesley, D. (2006). How the scale of effects on internationally designated nature conservation sites in Britain has been considered in decision making: a review of authoritative decisions. *English Nature Research Report*, No. 704. Peterborough: English Nature.

Howsam, P. (2003). The current status and impact of water quality law in Britain with particular reference to the Water Framework Directive. In *Managing our Aquatic Environment in the 21st Century: Contemporary Issues of Water Quality*, eds. C. Neal & I. Littlewood. London: British Hydrological Society Occasional Paper, pp. 1–7.

Huntington, C. W. & Sommarstrom, S. (2000). *An Evaluation of Selected Watershed Councils in the Pacific Northwest and Northern California*. Parts I, II, III. Prepared for Trout Unlimited and Pacific Rivers Council, Eugene, OR.

IUCN (2006). *Summary Statistics for Globally Threatened Species*. Gland: IUCN World Conservation Press.

Jacobsen, D. (2004). Contrasting patterns in local and zonal family richness of stream invertebrates along an Andean altitudinal gradient. *Freshwater Biology*, **49**, 1293–305.

Karr, J. R. & Chu, E. W. (2000). Sustaining living rivers. *Hydrobiologia*, **422/423**, 1–14.

Karr, J. R. & Yoder, C. O. (2004). Biological assessment and criteria improve total maximum daily load decision making. *Journal of Environmental Engineering*, **130**, 594–604.

Lanoo, M. (ed.) (2005). *Amphibian Declines: The Conservation Status of United States Species*. Berkeley, CA: University of California Press.

Lawler, J. J., Campbell, S. P., Guerry, A. D. *et al.* (2002). The scope and treatment of threats in recovery plans. *Ecological Applications*, **12**, 663–7.

Lowe, W. H. & Likens G. E. (2005). Moving headwater streams to the head of the class *BioScience*, **55**, 196–7.

Moyle, P. B. & Yoshiyama, R. M. (1994). Protection of aquatic biodiversity in California: a five-tiered approach. *Fisheries*, **19**, 6–18.

Park, K. (2004). Assessment and management of invasive alien predators. *Ecology and Society*, **9**, 12.

Perrings, C. (2002). Biological invasions in aquatic systems: the economic problem. *Bulletin of Marine Science*, **70**, 541–52.

Piper, J. (2002). Cumulative effects assessment and sustainable development: evidence from UK case studies. *EIA Review*, **22**, 17–36.

Revenga, C., Campbell, I., Abell, R., De Villiers, P. & Bryer, M. (2005). Prospects for monitoring freshwater ecosystems towards the 2010 targets. *Philosophical Transactions of the Royal Society of London. Series B, Biological Sciences*, **360**, 397–413.

Rieman, B. R., Lee, D. C. & Thurow, R. F. (1997). Distribution, status, and likely future trends of bull trout within the Columbia River and Klamath River Basins. *North American Journal of Fisheries Management*, **17**, 1111–25.

Schindler, D. W. (1987). Detecting ecosystem response to anthropogenic stress. *Canadian Journal of Fisheries and Aquatic Sciences*, **44**, 6–25.

Sedell, J. R., Reeves, G. H., Hauer, F. R., Stanford, J. A. & Hawkins, C. P. (1990). Role of refugia in recovery from disturbances: modern fragmented and disconnected river systems. *Environmental Management*, **14**, 711–24.

Smith, K. F., Sax, D. F. & Lafferty, K. D. (2006). Evidence for the role of infectious disease in species extinction and endangerment. *Conservation Biology*, **20**, 1349–57.

Spencer, C. N., McClelland, B. R. & Stanford, J. A. (1991). Shrimp stocking, salmon collapse, and eagle displacement: cascading interactions in the food web of a large aquatic ecosystem. *BioScience*, **41**, 14–21.

Stanford, J. A. & Ward, J. V. (1992). Management of aquatic resources in large catchments: recognizing interactions between ecosystem connectivity and environmental disturbance. In *Watershed Management*, ed. R. J. Naiman. New York: Springer-Verlag, pp. 91–124.

Stuart, S. N., Chanson, J. S., Cox, N. A. *et al.* (2004). Status and trends of amphibian declines and extinctions worldwide. *Science*, **306**, 1783–6.

Thomas, C. D., Cameron, A., Green, R. E. *et al.* (2003). Extinction risk from climate change. *Nature*, **427**, 145–8.

Trombulak, S. C. & Frissell, C. A. (2000). Review of ecological effects of roads on terrestrial and aquatic communities. *Conservation Biology*, **14**, 18–30.

Warren, M. L. & Burr, B. M. (1994). Status of freshwater fishes of the United States: overview of an imperiled fauna. *Fisheries*, **19**, 6–18.

Williams, J. D., Warren, M. L., Cummings, K. S., Harris, J. L. & Neves, R. J. (1993). Conservation of freshwater mussels of the United States and Canada. *Fisheries*, **18**, 5–22.

Williams, J. E. (1993). Preserves and refuges for native western fishes: history and management. In *Battle Against Extinction: Native Fish Management in the American West*, eds. W. L. Minckley and J. E. Deacon. Tucson: University of Arizona Press, pp. 171–89.

Williams, J. E., Johnson, J. E., Hendrickson, D. A. *et al.* (1989). Fishes of North America – endangered, threatened, or of special concern: 1989. *Fisheries*, **14**, 2–20.

Ziemer, R. R., Lewis, J., Rice, R. M. & Lisle, T. E. (1991). Modeling the cumulative watershed effects of forest management strategies. *Journal of Environmental Quality*, **20**, 36–42.

Zwick, P. (1992). Stream habitat fragmentation – a threat to biodiversity. *Biodiversity and Conservation*, **1**, 80–97.

6 · *Evaluating restoration potential*

T. E. L. LANGFORD AND C. A. FRISSELL

The music of the stream is caused by obstruction.

(p. 11, 'A life is too short' by Nicholas Fairbairn (1989).
Glasgow: Fontana, Collins)

Background

The art of the possible

Freshwater habitats have been modified and used for water supply, generation of power, movement of people and goods, for food and for disposal of wastes over many years. The restoration of such habitats to some pre-exploited state which constrains their uses by humans is, therefore, seldom universally socially acceptable. Because of the potential involvement of many other interests and users involved, particularly in managing river ecosystems (Prato, 2003), the potential for restoration of any lentic or lotic habitat must be evaluated using a wide variety of criteria of which ecological conservation may have low priority. In practice, therefore, selecting a location for 'restoration' is a complex procedure comprising a mixture of quantifiable criteria and subjective, intuitive decisions based on experience or interpretations of theory (Golet *et al.*, 2006). While aesthetics, tradition, equity of human uses and political concerns play major roles in defining socially desirable consequences, ecological and physical sciences continue to play a central role in defining the possible.

This chapter examines the principles of river and lake restoration from the viewpoint of ecology and conservation. Although there are some differences in approach between the UK and the USA, there are many similarities. Thus, the text that follows, while providing examples to illustrate differences, is more a joint examination of how things work in both countries now, and how they should work in future if habitat restoration is to be an effective tool in freshwater conservation.

The scales and aims of potential restoration projects vary considerably from small-scale, low-intrusion schemes (Cowx & Welcomme, 1998), to

Assessing the Conservation Value of Fresh Waters, ed. Philip J. Boon and Catherine M. Pringle. Published by Cambridge University Press.

large-scale, multi-purpose, multi-faceted schemes using large machinery on large river systems or lakes (Tockner *et al.*, 1998; Astrack, 2005). The selection of sites may be based on many criteria other than conservation, but the total objectives should include strategies to conserve and, wherever possible, enhance whatever component of the ecosystem is regarded as natural or ecologically valuable (Boon *et al.*, 2002) and also to allow the potential reinstatement of natural components which had existed prior to human impacts.

River restoration has become more prevalent in Europe since flood damage cost some 30 000 million dollars around the turn of the twentieth century (Pearce, 2006). Huge payouts by the insurance industry and governments in several developed regions of the world, together with the damaging local and national political embarrassment caused by drastic inundation of badly planned floodplain development, led to restoration projects on various scales. Although many of these were ostensibly aimed at ecological or aesthetic objectives (Ormerod, 2004), they were, in truth, an attempt to redress the mistakes of earlier planning, political and engineering decisions. If ecological benefits could be identified this served to enhance the acceptability of a project. Similar circumstances also prevailed in recent decades in the USA after floods on the Mississippi and other floodplain rivers. The importance of restoration to flood prevention is indicated in the UK by the fact that in some regions restoration projects are funded by the flood defence budget of the Environment Agency (Skinner & Bruce-Burgess, 2005). Between 1999 and 2003, projects increased by 150%. Thus, whilst the choice of sites may seem to ecologists somewhat ad hoc or even potentially damaging, they may be more logical for flood management. The need for scrutiny of restoration schemes that might negatively affect species of importance or protected communities is paramount.

Overall ecosystem change has clearly been a secondary or even peripheral consideration in many schemes, despite their stated ecological or conservation credentials. For example, in the UK, the planning and design of some 42% of 538 restoration schemes before 2001 involved in part 'improvement' of either fisheries or ecological functions or assemblages (Clarke *et al.*, 2003). The remainder presumably mostly involved flood alleviation, flow restoration or aesthetics. Ormerod (2004) noted that many published papers on restoration used the word 'habitat' in their texts which implied ecological considerations, but relatively few papers assessing the effects of restoration on rivers or lakes contain detailed, scientifically viable ecological data from either pre- or post-project studies.

Lake restoration, typically, through improvement of water quality, reduction of eutrophication and control of alien invasive species, has become more important as sport fishery and recreation demands have increased, demand for water supply has grown and lakeshore and catchment development has expanded and intensified (Welch & Cooke, 1987; Moore *et al.*, 2003).

Introducing complexity to selection criteria

In recent years, direct necessity, public safety, economic desirability, amenity, conservation and ecosystem enhancement have been used separately or collectively as reasons for restoration, but to evaluate the potential success or failure of any restoration project the purpose of the project needs to be defined clearly, preferably by some quantifiable or quantified methods. Ecologically, the purpose may simply be the direct reinstatement or conservation of a single species or the indirect re-introduction of some degree of biological diversity or function (Petts & Calow, 1996; Roni, 2005). In many cases the choice of sites is a fait accompli for societal reasons despite possible reservations on conservation or ecological grounds, and the prediction of ecological success is more complex.

The introduction of concepts such as 'ecosystem health' (Karr & Dudley, 1981; Maddock, 1999; An & Choi, 2003), 'ecosystem integrity' (Karr, 1991), 'ecological architecture' (Davis & Slobodkin, 2004), 'ecosystem services' (Cairns, 2000; Holmes *et al.*, 2004), the 'normative ecosystem' (Stanford *et al.*, 1996), 'ecological status' (2000/60/EC) and 'stakeholder' or 'socio-economic interests' (Clark, 2002) has increased the number of criteria by which the potential for restoration schemes may be judged to the point where it is almost impossible to oppose a proposed scheme or declare it a total success or failure once completed. Proponents of more recent restoration policies have moved away from ecologically defined objectives towards the satisfaction of human 'stakeholders' (Cairns, 2000; Clark, 2002). Many schemes are at least dual purpose, for example fisheries or biological diversity allied to flood alleviation, aesthetics and floodplain connectivity. Local residents may also be prepared to pay, through local taxes, towards aesthetic or conservation-based restoration schemes (Loomis *et al.*, 2000). The choice of ecological or conservation-based restoration, therefore, '*is only one of many possible choices*' (Davis & Slobodkin, 2004).

Concepts such as normative ecosystems (Stanford *et al.*, 1996) mix biological variables relevant to the habitat and native species requirements with variables relevant to the human community so that there may be cultural, aesthetic and socio-economic considerations involved. The normative river concept provides a theoretical framework by which a holistic restoration goal is defined. In this concept, choices of natural processes to be retained are balanced with the possibility that water users may require some acceptable deviation from the natural state. Such concepts clearly provide the basis for future restoration though conservation may be a lower priority than might be desired.

Floodplain connectivity and biological diversity, the holistic view of the river and its catchment (Hynes, 1975; Vannote *et al.*, 1980; Ward *et al.*, 2001) and wider supra-catchment considerations such as aerial deposition (Howells, 1990) add further complexities to the evaluation and selection of sites for restoration and conservation. Because of the complexity of the criteria, proponents and opponents of any particular scheme may 'balance' detrimental effects in one variable with beneficial effects in another, and thus justify or condemn the costs and physical disturbances as they wish. In this way schemes judged a restoration success in political terms are often not so in ecological terms (Palmer *et al.*, 2005). From the priority on which a project is based, targets and criteria for success can be determined, though it is likely that evaluation for conservation and restoration may be 'often a process that is more intuitive than scientific' (Usher, 1986).

Conservation as a basis of site selection

Aims of the ecological evaluation procedure

Given all the potential criteria and reasons for restoring a particular river or lake, evaluation of the consequences for conservation or ecological improvement has to set positive or negative predictions in the context of the overall objectives (Kondolf, 1995, 1998). Limits to benefits and possibilities of adverse effects should be considered and stated in the final proposals, but potential adverse effects are rarely stated (Kondolf, 1998; Langford *et al.*, 2000). Importantly, the pre-proposal evaluation requires an assessment of the relationship of a potential site to the larger environment including factors such as water and sediment run-off, thermal regimes of the catchment, upstream or aerial sources of important contaminants and population source pools of target species or threatening invasive species for potential colonization (Ebersole *et al.*, 1997; Frissell *et al.*, 2001; Langford *et al.*, in press). Predictions for sites considered for restoration also need to assess the theoretical and practical constraints to success. Such evaluation should also include the potential effects on conservation targets of a 'do nothing' or 'no action' alternative (Boon, 1998).

In pre-proposal and post-project evaluations there should be some quantifiable measurement which can be graded in relation to some quantified target (Usher, 1986). This is rarely the case in practice, particularly for pre-project evaluation, perhaps because this is the most difficult form of evaluation. To be effective there is a need for extensive predictions for which few specific data or quantitative models are available. Though there are many classic data in the literature that could be used as bases for prediction (Hynes, 1970; Petts & Calow, 1996), Frissell *et al.* (2001) postulate that crucial criteria for evaluation are efficiency and cost-effectiveness. Large benefits can accrue from limited actions, notably by ensuring the protection and preservation of those parts of ecosystems and ecosystem processes that remain relatively unaltered from their natural historical condition (Frissell & Bayles, 1996; Frissell *et al.*, 2001). This is a core consideration in the normative river concept (Stanford *et al.*, 1996).

Components of the selection process

The main components of the evaluation procedure for selecting potential waters for restoration are:

- establishing the need and the priority for the restoration;
- providing objectives (targets) to justify expenditure and resources;
- identifying the physical, biological and social causes of undesired and compromised ecosystem conditions, and ensuring that these are amenable to reversal through remedial action;
- assessing the probability of success of the interventions (short-term objectives) in overall ecological and conservation terms;
- assessing the sustainability of the restored condition (long-term);
- providing a framework and establishing the methods on which the post-restoration evaluation may be based.

Establishing priorities

The assumption prior to evaluation for restoration must be that the project is deemed to be necessary for some defined purpose or purposes which might include:

- direct human safety or welfare, i.e. the project will protect or conserve life or living conditions for human populations – e.g. flood relief, water supply;
- economic production, i.e. the project would be of financial advantage to the human population – e.g. commercial fishery, business or agriculture;

- amenity, i.e. the project would improve the pleasure and appreciation of the environment for human populations;
- ecology, conservation, i.e. the project is intended to maintain species protected by law or to protect and sustain biological diversity, naturalness or some other set of ecological criteria.

There is often no statutory requirement for all projects labelled as 'restoration' to go through formal ecological evaluation either before or after implementation in either the USA or the UK. Exceptions in the USA are on federal lands in large government and privately funded projects where the National Environmental Policy Act and some similar state laws do require comprehensive evaluation and disclosure of the human and environmental effects (including ecological effects) of major projects, though smaller projects may escape such scrutiny in both the USA and the UK.

Objectives for projects are often loosely defined – for example, 'to improve biodiversity' or 'to reduce flood-risk' in particular – and there is seldom a legal requirement in the UK for quantifiable objectives which proponents of a scheme are obliged to meet and therefore no legal penalties if objectives are not reached. However, in the USA some rights secured by Native American tribes under treaty law, and many actions undertaken to recover species protected under the US Endangered Species Act, do include explicit targets and numerical criteria, and there are at least potential consequences to individuals, businesses and government when these targets are not met.

Defining achievable ecological targets for conservation

It is clear that most sites for restoration are not now selected on the basis of conservation alone or even in many cases with conservation as a primary criterion. Thus, defining conservation targets for any scheme must fit an overall project target and may have to be defined within an overall priority context both before and after project implementation.

Setting realistic targets for biological conservation within an overall restoration project requires

- an understanding of ecological theory;
- an understanding of physical and chemical processes in rivers and lakes;
- an analysis of relevant (but not necessarily restoration) literature;
- data from case histories where possible;
- a prediction of potential adverse effects of the restoration activities;
- some subjective, intuitive instinct that cannot be readily quantified.

Because of other priorities the conservation target may have to be less than ideal, a condition accepted by the framework of the 'normative state concept' (Stanford *et al.*, 1996).

There is no universally applicable model of a river or lake ecosystem, even though classification systems attempt to delimit types based on regional reference sites (Wright *et al.*, 2000; Moss *et al.*, 2003). Classifications can identify affiliation based on generic or underlying biophysical templates that set some broad scope for ecosystem behaviour and performance (Frissell *et al.*, 1986), but imposed on this template is a high degree of diversity together with local spatial and temporal individuality in both human alteration and natural disturbances. Thus, routine application of generic methods of restoration without regard to local factors may be a recipe for frequent failure. Targets must inevitably be individual for a set ecosystem, species or community (Larsen, 1996), though they may be based on other habitats considered or assessed as 'typical' or 'natural' in that locality or region (Usher, 1986; Boon *et al.*, 2002).

The principle of ecological target setting, whether at the ecosystem level or for conserving a single species, is the same for running and standing waters, namely

- What biological system are we trying to restore?
- Will any species targeted for conservation survive within the altered ecosystem created by the restoration?

Because any restoration activity usually involves either physical or chemical disturbance of the habitat, the conservation of any target species must be assessed within the context of potential ecosystem change. Ecological targets can be

- taxonomically defined either as some form of diversity or species richness value;
- defined by an ecological status measure such as biotic indices or scores;
- defined by the need to conserve or encourage species or habitat-type;
- functionally defined as processes or relative abundance of target functional groups, e.g. grazers, shredders, filterers;
- defined by a target abundance or productivity value for specific groups, e.g. fish, algae;
- defined by some method of comparing communities, e.g. ordination or similarity analysis;
- defined by a pragmatic target such as fish catches or spawning success, though these may also be regarded as economic or societal targets.

The target can be qualitative or quantitative. It can be set by reconstruction of previous ecosystems (Battarbee, 2000) or by comparison with reference sites (Wright *et al.*, 2000). The target model might be some version of a previous existence or possibly a view of a future, very different habitat that bears little relation to the past or present. There are now many definitions of restoration that might be used as a basis for setting targets ranging from total objectives such as 'the process of returning (*a habitat*) to a condition that promotes re-expression of natural ecosystem structure and function' (Moerke & Lamberti, 2004) to specific objectives such as the protection or reinstatement of 'key species, the communities of which they are part and the ecological functions they provide' (Ormerod, 2004).

The question of whether it is more important to conserve or restore ecosystem functioning or taxonomic diversity is the subject of debate. Functional targets will depend on measurements of primary or secondary production, and key functional linkages will be related to the structure of the food-web (Palmer *et al.*, 1997). Ultimately, however, this food web-function target will depend on certain species being present and the physical and chemical habitat for these being produced. Thus the functional model could well be defined using taxonomic data. Targets based on eco-hydraulic models for fish such as PHABSIM (Bovee, 1982), or invertebrates (Kemp *et al.*, 2000; Bockelmann *et al.*, 2004) have played a role in ecological restoration with variable success, although alternative, more integrative approaches to flow restoration are embodied in the 'normative rivers concept'.

Targets for the conservation of single species or communities in rivers based on chemical rehabilitation may be simpler to establish, though published models are rare (Langford *et al.*, in press) (Figure 6.1). Such generic models are not readily available for relationships between physical structure and communities, though there are many descriptions and tentative qualitative models of relationships between physical habitat and fish species from various regions (Milner *et al.*, 1985; Heggenes, 1988). Targets for the conservation of species such as migratory fish in rivers as a result of restoring connectivity may be based on historical data or on catch statistics.

The success or failure of any project to attain a target may depend on the scale as well as the ecosystem context of the project (Frissell & Nawa, 1992). Most projects with an ecological or fishery target are carried out at the reach scale with a view to benefit on the larger scale. The increased abundance of a target species by creating specific refugia or the enhancement of its life cycle by creating spawning and nursery habitats are fairly

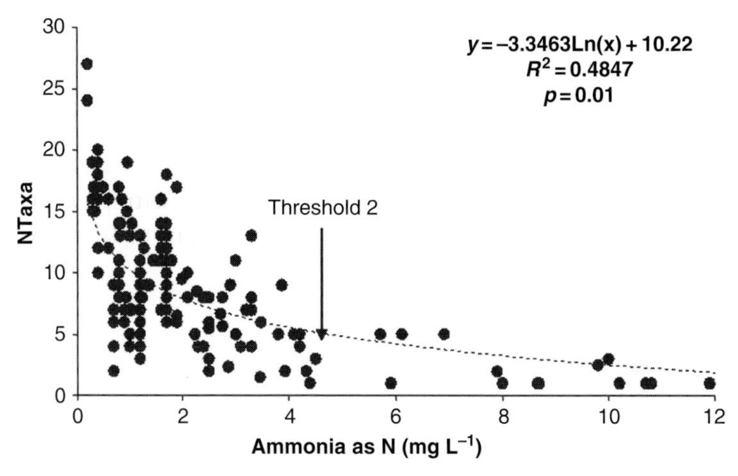

Figure 6.1. Relationship between numbers of taxa occurring (NTaxa) in relation to annual mean ammonia (as N) concentration. Data from three locations on three UK rivers. After Langford *et al.* (in press). Threshold 2 indicates the point where less tolerant species begin to colonize rapidly, although there are variations with year and location.

common objectives, particularly for fishery improvements. However, the manipulation of a channel to provide, for example, an altered ratio of pool to riffle habitat may have different effects on different species or ontogenic stages because of habitat preferences (Langford, 2006). Increasing physical diversity within a water body by using woody debris or by substrate manipulation and augmentation will not necessarily result in an increase in overall biological diversity (Madsen *et al.*, 2005) if the community is limited by water chemistry (e.g. naturally low mineral content, acidity or pollution) or by sedimentation and channel instability stemming from upstream or catchment-wide land use (Frissell & Nawa, 1992).

Setting conservation or overall ecological targets for lake restoration (Janse & van Liere, 1995) can be simpler than for rivers in that the problems are mainly chemical or involve invasive species. Thus, whether nutrient loads are reduced by eliminating inputs (Diaz-Pardo *et al.*, 1998), by lake flushing (Hosper, 1998) or by removing sediments (van Duin *et al.*, 1998), there are mathematical models relating nutrient concentrations to algal biomass and chlorophyll *a* concentrations that can be used to set targets. Models based on macrophyte biomass may, however, be more problematic (Van Nes *et al.*, 2002) in that achieving some optimal condition can depend on factors such as setting accurate harvesting targets often involving unrealistic costs. Targets based on the reconstruction of previous ecosystems or communities can benefit from palaeo-ecological reconstruction using the hard parts of organisms deposited in bottom sediments, including crustaceans, ostracods, diatoms and chironomids (Battarbee, 2000). However, there may be irreversible components in lakes because of changes in trophic chains or sediment storage of nutrients or toxins. Successful restoration, for example by removal of unwanted or alien fish, was only possible in one lake because the other elements of the ecosystem were intact (Knapp *et al.*, 2001).

Myth and uncertainty

Unsatisfactory restoration schemes, ostensibly those that do not approach or maintain the objectives (often unquantified in any case) originate from a 'failure to recognize and address uncertainty' (Hilderbrand *et al.*, 2005). The 'field of dreams' myth expects that re-creating a physical or chemical structure in a habitat will result in predictable changes in community structure or species abundance. Although physical and chemical structure may set a template for such a process (Frissell *et al.*, 1986), this simple expectation is complicated by scaled and nested biological processes such as mobility and availability of organisms for re-colonization and stochasticity of community composition. Introducing species may also not produce the desired objective as the physical or chemical changes in the habitat may not be sustainable, particularly if there are scalar problems (Frissell & Nawa, 1992). Finally, the expectation that human intervention can control other ecosystems within some expected and fairly narrow range of variability is almost always disappointed (Frissell & Bayles, 1996). Despite the myths, there are aspects of ecosystems over which human intervention can exercise some control, but the element of uncertainty always remains and should be scheduled into any evaluations for restoration sites. In addition, recognition of the myths and of the uncertainties can be valuable in setting targets and in evaluating potential restoration sites (Frissell & Bayles, 1996; Frissell & Ralph, 1998; Ralph & Poole, 2003).

A framework for evaluating restoration potential

Evaluation protocols for conservation

Composite and comprehensive methodologies are now available for the evaluation of aquatic habitats for conservation, and modifications allow the evaluation of restoration potential and restoration success (Frissell, 1997; River Restoration Centre, 1999; Xu *et al.*, 2001; Boon *et al.*, 2002; Society for Ecological Restoration, 2002; Moss *et al.*, 2003). One of the most recently developed and modified protocols for evaluating rivers in the UK for conservation is the SERCON (System for Evaluating Rivers for Conservation) procedure (see Chapter 7; Boon, 1992; Boon *et al.*, 1998, 2002). Assessment of the ecological status of lakes in Europe may be based on the ECOFRAME methodology (Moss *et al.*, 2003) or on a system of comparing ecological models with actual observations (Xu *et al.*, 2001).

SERCON 1 (Boon *et al.*, 1998) was initially designed to assess river habitats for conservation, in terms of generally accepted conservation criteria such as 'naturalness', 'representativeness' and 'rarity'. SERCON 2 (Boon *et al.*, 2002) contains an additional suite of specific applications, including one on river rehabilitation, where the decision-making process relates to sites, reaches or larger-scale units for potential restoration. Where the scores for environmental impacts vary widely there may be a good case for restoration, as reducing a few heavy impacts may be highly effective in improving habitat and ecological quality. The ultimate choice will, however, depend on the priorities of the project and the ecological targets set. With relatively little modification the SERCON procedure could be used for evaluation of lakes for restoration. However, SERCON and its applications are not yet, unfortunately, used routinely and choosing sites for restoration is still mostly made on subjective, often politically expedient bases.

In the UK, the River Restoration Centre (RRC) (www.therrc.co,uk), formed in 1998 has been a national centre for the coordination and dissemination of information on river restoration projects and has produced valuable guidelines for project planning and implementation (RRC, 1999, 2002). Unfortunately, very little of the literature or proceedings of conferences have been focused on evaluating the potential for restoration or even, until relatively recently, on the negative ecological aspects of restoration schemes.

In North America, less formal and analytical methods are used for setting stream restoration priorities. Most choices are based on an assessment of the importance of habitat patches for fish, mussels or other native species that are either protected under the US Endangered Species Act, or are declining, but increasing production of abundant commercially exploited fish, aesthetics and access to sites for educational purposes, are also frequent considerations. In most cases local or regional political will is paramount. Even when there are formally designated priority areas for restoration efforts, such as the so-called 'Key Watersheds' designated on federal forest lands in the Pacific Northwest states (Forest Ecosystem

Management Assessment Team (FEMAT), 1993), actual accomplishment of the work is dependent on local political and agency effort. As in the UK, manuals and guides for implementing restoration projects are widely available (Roni, 2005), but again they concentrate more on post-project assessment rather than site selection and pre-project evaluation.

The selection of lakes for restoration has concentrated more on chemical rehabilitation (particularly the reduction of eutrophication) and removal of alien or invasive species than on physical modification or biological conservation (Welch & Cooke, 1987). Recreation, fisheries and water supply are generally high-priority objectives. The establishment of 'reference' conditions for European lakes is problematic (Moss *et al.*, 2003; Irvine, 2004). For example, the largest exercise to classify shallow lake types using 66 lakes and 28 physical, chemical and biological parameters found that few were classified as good or high status despite minimal human interference (Moss *et al.*, 2003). There is as yet little agreement on the criteria that such lakes need to meet to reach the highest classifications though the choice of indicators of ecological status or health (Xu *et al.*, 2001) should be related to the objectives or targets for the restoration or rehabilitation scheme.

The effects of reducing eutrophication, usually by reducing inputs of phosphorus (van Duin *et al.*,1998) show most clearly in the reductions of blue-green algae and chlorophyll concentrations. In Dutch lakes (Klinge *et al.*, 1995), the relative dominance of piscivorous fish to prey fish was considered as a useful indicator of the change in eutrophic status and was one aspect of a framework methodology. Bain *et al.* (2000) listed nine 'structure' indicators used to 'identify impaired biological integrity of lakes, to measure restoration responses … and to test the feasibility of restoring biological integrity through fish-community manipulation'. These included taxonomic, functional and multi-metric indices together with flow, chemistry, physical structure and riparian characteristics.

Despite macro-invertebrates and fish being included in the relevant EC Directive as indicators of lake condition they are impractically expensive to sample comprehensively and require very specific sampling equipment for the range of physical conditions encountered (Moss *et al.*, 2003). Restricted habitat sampling or surrogate species as indicators (Briers & Biggs, 2003) may be useful alternatives. In the USA, lakes are also targeted for restoration based largely on local, social and political considerations. A few lakes have been targeted for biological restoration of native species communities, e.g. by eliminating introduced fish with the intent of restoring indigenous populations of frogs or salamanders that are vulnerable to fish predation (Knapp *et al.*, 2001). However, Drake and Naiman (2000) found that lakes in Mt. Rainier National Park were not ecologically restored by removing introduced trout, illustrating that the effects of introduced fish on ecological integrity can be long-lasting or possibly permanent in some respects.

Predicting the consequences of restoration on conservation and ecology

The complexity of physical, chemical and biological processes in time and space in small-scale habitats will preclude certainty of predictions, particularly at larger scales (Kondolf, 1995; Hilderbrand *et al.*, 2005), and the best prediction can only be of some degree of movement towards a set target to produce a system that is relatively stable in the longer term. For single-species conservation, it may be relatively easy to measure the changes in abundance, but for species such as migratory fish the reasons for changes in abundance may relate to stages outside the restored environment and this must be considered when assessing potential success of a project. Indicators of overall ecological change are a matter of choice but clearly should be relevant to the stated objectives of the restoration itself. The success or failure of any scheme can depend on many factors but there is 'little agreement ... on what constitutes a successful river (*or lake*) restoration effort' (Palmer *et al.*, 2005).

Restoration as disturbance and effects on conservation

The theoretical concepts of disturbance and its aftermath, together with empirical data from disturbance studies, could form the basis for predicting consequences of restoration activities on communities or the conservation of single species (Jordan *et al.*, 1987; Milner, 1996; Winterhalder *et al.*, 2004). Disturbance has been defined as 'a perturbation which is a relatively discrete event in time that removes organisms and opens up space or other resources' (Hildrew & Giller, 1994). Restoration activities may therefore be defined as 'the use of an artificial disturbance to attempt to direct the effects of a previous disturbance towards some designated state that is perceived to be more desirable' though the most objective definition is 'a disturbance designed to alter the effects of a previous disturbance and superimposed upon it' (Langford, 2006).

Any restoration involves some physical or chemical manipulation of the habitat and hence of the ecosystem. Therefore, to evaluate the potential for restoration or rehabilitation of any water body it is necessary to understand the theoretical aspects of the physical, chemical and biological processes involved in its development and maintenance and their scale of operation. For example, reinstating sinuosity in a river by excavating and realigning channels may be directly beneficial to humans because of flood alleviation or aesthetics. From the perspective of the organisms resident in the channel, however, the excavations involved would be a major disturbance which may be as destructive in the short term as the disturbance that altered the natural habitat in the first place (Hildrew & Giller, 1994). Ecological theory dictates that such habitat alteration is likely to benefit some species at the expense of others mainly because of factors such as habitat preference, resource partitioning, niche separation and competitive exclusion. Species targeted for conservation can thus be losers or winners during restoration.

Exploitation disturbances such as abstraction, embankment and canalization of river channels, or pollution of rivers and lakes, can virtually eliminate the natural flora and fauna and replace them with alternative communities (Hynes, 1960; Moss, 1998). However, natural disturbances such as floods, landslips or dramatic lateral movements of river channels and erosion can also be on vast scales and can create similar conditions to artificial disturbances (Frissell *et al.*, 1986). Data from studies of such natural phenomena may, therefore, be of use in predicting the effects of artificial restoration. The major difference between natural disturbances and restoration disturbance is that the former can occur at different frequencies over time whereas the latter usually occur only once or twice.

Recovery after restoration: implications for conservation

The ecological processes following restoration will depend on the degree of establishment of the necessary physical and chemical processes but will be similar to recovery after natural disturbance, namely, re-colonization, succession and stabilization (Milner, 1996).

Evaluating the potential consequences of restoration on species or communities targeted for conservation involves measures of the rate and extent of post-disturbance recovery or re-colonization by a target species (Milner, 1996). The rate of recovery in both standing and running waters is dependent upon

- the severity of the exploitation disturbance in relation to the resistance and resilience of the ecosystem and its components;
- the severity of the restoration disturbance in relation to the resilience and resistance of the established ecosystem;

- the suitability of the physical and chemical conditions after restoration for the target ecosystem;
- the proximity of potential sources of species for re-colonization;
- the mobility of the relevant species;
- the mechanism of re-colonization, succession and stabilization;
- events outside the restored habitat, such as the effects of sea conditions for migratory fish species that can affect abundance of freshwater stages (Kondolf, 1995).

Scale and intensity of restoration disturbance

The scale and intensity of any restoration scheme will be a major factor in the recovery of the ecosystem, and the conservation and sustainability of target species populations (Figure 6.2). Although some of the evaluation methods for rivers are based on reach-scale assessments or catchment sections (Boon *et al.*, 1998, 2002), there is no universal ecologically acceptable unit for restoration.

Evaluating the recovery of species populations or communities targeted for conservation is generally easier for smaller scales than larger scales usually because of the proximity of reference or control sites and a good knowledge of mechanisms of recovery from disturbance. At larger scales

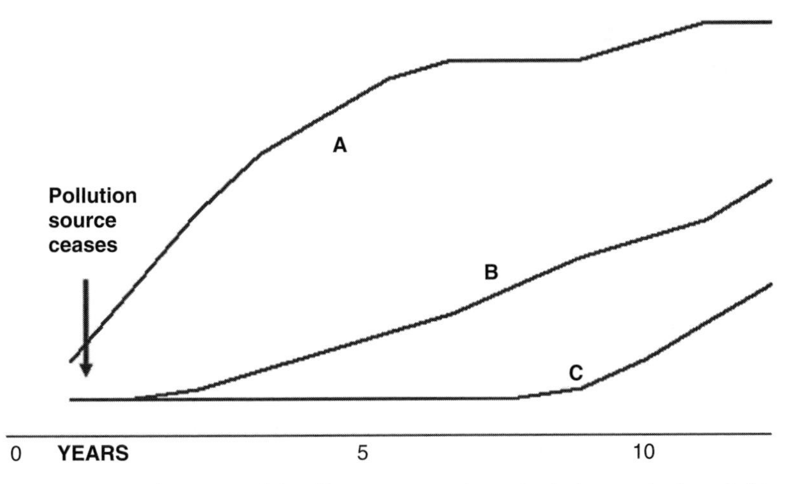

Figure 6.2. Qualitative models of improvement in ecological status in three habitat types following cessation of a pollution source. A. Clean water, sources of species for re-colonization close and very mobile; B. Moderately clean water, sources of species not so close and species not so mobile; C. Water quality much improved, sources of biota very inaccessible. Models based on empirical data from three UK rivers.

it is difficult, not only because of the greater numbers of variables involved but also because of a lack of reference habitats at the same scale. Restoration on a larger scale in rivers – for example, the destruction of artificial obstacles to migration (e.g. dams, weirs, locks) (Hart *et al.*, 2002) or the recovery of a polluted reach (Langford *et al.*, in press) – can result in very rapid re-colonization by migratory species. Where biota are less available for recolonization, the process of recovery may be very slow (Figure 6.3). Clearly, the success of restoration to enhance the conservation of a target species will also depend upon the availability of that species and its mobility.

Models or indicators of recovery potential may come from pristine sites such as large undeveloped regions in western North America (Sedell *et al.*, 1990; Frissell & Bayles, 1996). Restoration becomes more uncertain and less significant in its potential success when smaller fragments of catchments, lakes and streams are the only pieces of ecosystems that remain as suitable candidates for restoration. Such pristine sites are rarely available in the UK where most rivers and standing waters are affected by human

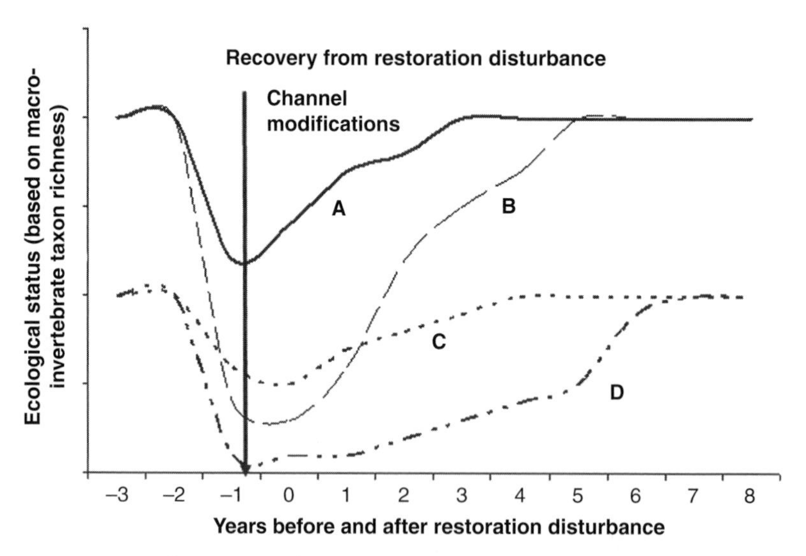

Figure 6.3. Qualitative models of recovery rates of macro-invertebrate communities in streams with various degrees of restoration disturbance. A. Chalk stream, (high energy) low disturbance, high resilience; B. Chalk stream (high energy) high disturbance, high resilience; C. Acid stream (low energy), low disturbance, low resilience; D. Acid stream (low energy) high disturbance, low resilience. Models based on data from UK streams.

activity to some degree, and, although reference communities are used to assess human impacts (Wright *et al.*, 2000), it is recognized that even these may exhibit some degree of human impact.

Learning from success and failure: the two impostors

Projects judged as successful or as failed can provide valuable data for site selection. The use of case-history data can, however, be problematic, mainly because of the equivocal conclusions from the relatively few post-project studies carried out. While it is clear for studies in the USA that many projects have failed to sustain both physical and ecological objectives (Frissell & Nawa, 1992), others claim considerable success (House, 1996). One of the main reasons for failure has been where relatively small-scale projects (e.g. reach) have been overwhelmed by processes at a larger scale (e.g. catchment) which had not been predicted. For example, restoring meanders in a stream released some 78 tonnes of sediment obliterating riffle faunas and macrophytes downstream for about two years (Friberg *et al.*, 1998). Similarly, in a small Wisconsin river, the removal of a dam caused a 30% decline in the density of unionid mussels downstream as a result of silt deposition (Sethi *et al.*, 2004). Upstream, the mussels suffered up to 95% mortality as a result of stranding and desiccation in the drained reservoir. During restoration one rare species was lost. At the same time increased connectivity and fish passage improved the migratory fish populations.

In some English chalk streams, channel restoration and riparian fencing made little difference to the diversity of in-stream vegetation but in the ungrazed, fenced reaches of one stream there were eight fewer riparian and bankside macrophyte species than in the unfenced, grazed reaches (Langford *et al.*, 2000). However, intensive cattle grazing in arid land streams of the American West appears to have precisely the opposite effect, simplifying riparian vegetation and channel structure and their dependent biotic communities. In forested streams, the removal of riparian cover during restoration can cause increases in water temperature that may exceed limits tolerable by species that can exist in cooler waters (Langford, 2006) (Figure 6.4). Thermal characteristics of natural and impounded lakes and streams impose similarly strict physiological limits on performance for most species, such as coldwater fish (see Langford, 1983, 1990), and future climate change may shift the attainable thermal regime so that historical conditions are difficult to restore.

Figure 6.4. A case for protection from restoration. This new channel was dug in a trout stream system as part of a LIFE 3 restoration scheme in the New Forest in Southern England. The previous stream had been straightened in the 1950s but was fully shaded with a gravel bed. The channel shown here was dug through bare clay and trees clear-felled with no replanting. The channel was engineered with trapezoidal section. Water temperatures on hot days reached up to 27°C. The primary stated aims of the project were to encourage alluvial forest and mitigate flooding downstream. To raise the stream bed and encourage overbank flow, tonnes of gravel, sand and clay mixture (hoggin) were introduced. The fines were transported downstream in spates, smothering riffles outside the restored sections. Iron in the added mixture caused growths of iron bacteria and extensive flocs of ferric hydroxide on the stream bed smothering both invertebrates and small riffle fish. Despite these problems, the channel was colonized by fish when flows were adequate some two years after completion.

Future evaluation for restoration and conservation

Clearly, the decisions on where to implement river and lake restoration schemes have become more difficult over the past decade as more and varied criteria have been defined. Ecological decisions can be clarified by using a semi-quantitative procedure such as SERCON 2 (Boon *et al.*, 2002), but the totality of the decision-making process now requires much more information synthesis and analysis (Riechert *et al.*, 2007; Woolsey *et al.*, 2007). As a generally applicable concept the 'normative state' is a suitable formal basis on which selection may be based. From this, the acceptance by ecologists that ecology and conservation are, in many cases, minor considerations in priorities for restoration, may be the key to more acknowledgement from non-ecologists that schemes could actually damage ecosystems if poorly planned and implemented.

We have not explored here the detailed relationships between legislation, restoration and conservation mainly because we aimed to examine the practical complexities of predicting and evaluating sites for restoration and conservation. However, legislation has already played a vital role in the rehabilitation and protection of aquatic ecosystems. The most effective legislation has been that aimed at pollution control, in particular the Rivers (Prevention of Pollution) Act (1961) in the UK and the Clean Water Act (1972) in the USA. The 1961 Act was the first Act of Parliament to establish retrospective controls on effluents in the UK and was almost immediately effective, though new technologies, new infrastructure and the change from heavy to lighter industry were also significant factors. Subsequent Acts such as the Control of Pollution Act (1974) and the Wildlife and Countryside Act (1981) strengthened the requirements for clean water and for the consideration of ecology. The Endangered Species Act in the USA for the first time protected species on the brink of local or national extinction. In both the UK and the USA, rivers which were once foul, noxious and fishless for miles are now clean with diverse flora and fauna. Any improvements from now on are likely to be, for the most part, relatively small in the UK and the USA compared with those of the last 40 years. Much effort will probably be aimed at urban streams.

The role of the European Water Framework Directive (WFD) (2000/60/EC) and the EC Habitats Directive (92/43/EEC) (Chapter 2) as agents for better conservation and ecological quality may, arguably, be less effective in the UK than purely national legislation. For example, although ecological quality and ecological status are explicitly used as criteria in the WFD, there are very poor definitions of the terms. Furthermore, the classification of water bodies as 'heavily modified' as a reason for setting lower ecological standards may provide similar escape routes for avoiding improvement or control measures as were included in UK Acts of Parliament before 1961. The attention now given under the WFD to 'hydromorphology' for surface waters is perhaps somewhat overdue, but the improvements in ecology as a result of physical restoration in the UK are likely to be less significant overall than from pollution control. Even so, there are some species, and some urban and upland streams, that will benefit from physical change. In the USA, improvements in hydromorphology may be more beneficial, particularly where streams in upland areas have been heavily degraded by mining or forestry operations.

There is clearly a need for more research, specifically into the direct and indirect relationships between the physical nature of rivers and lakes and ecosystem structure and functioning. At the same time, comprehensive literature reviews, particularly of post-project data, could provide the predictive background from which models could be proposed for evaluating ecological restoration potential. In future, projects will need to be based on catchment or even supra-catchment scales in some regions. The study of the more pristine habitats of the world can also provide data for evaluations (Ward *et al.*, 2001). Despite all these sources of information and the plethora of literature proposing methods and approaches, it is likely that most large-scale freshwater 'restoration' projects in the near future will be proposed and implemented on the basis of political expediency or public safety rather than on the grounds of conservation or ecology. The role of ecologists will be to ensure that native flora and fauna, and the places they inhabit, are conserved as far as possible and protected from the consequences both of exploitation and restoration.

Acknowledgements

The authors would like to thank Professor Phil Boon and Dr Cathy Pringle for their help, advice and patience. Also, Jean Langford compiled the bibliography and copy edited the text several times, and our colleagues offered advice and help for which we are grateful. Pacific Rivers Council supported the second author in the preparation of this chapter.

References

An, K. G. & Choi, S. S. (2003). An assessment of aquatic ecosystem health in a temperate watershed using the index of biological integrity. *Journal of Environmental Science and Health*, **38**, 1115–30.

Astrack, R. F. (2005). The upper Mississippi river comprehensive flood damage reduction study: taking a fresh look at an old problem on a basin-wide level. *Journal of Contemporary Water Research & Education*, **130**, 31–5.

Bain, M. B., Harig, A. L., Loucks, D. P., Goforth, R. R. & Mills, K. E. (2000). Aquatic ecosystem protection and restoration: advances in methods for assessment and evaluation. *Environmental Science & Policy*, **3**, 589–98.

Battarbee, R. W. (2000). Palaeolimnological approaches to climate change, with special regard to the biological record. *Quaternary Science Reviews*, **19**, 107–24.

Bockelmann, B. N., Fenrich, E. K., Lin, B. & Falconer, R. A. (2004). Development of an ecohydraulics model for stream and river restoration. *Ecological Engineering*, **22**, 227–35.

Boon, P. J. (1992). Essential elements in the case for river conservation. In *River Conservation and Management*, eds. P. J. Boon, P. Calow & G. E. Petts. Chichester: John Wiley, pp. 11–33.

Boon, P. J. (1998). River restoration in five dimensions. *Aquatic Conservation: Marine & Freshwater Ecosystems*, **8**, 257–64.

Boon, P. J., Wilkinson, J. & Martin, J. (1998). The application of SERCON (System for Evaluating Rivers for Conservation) to a selection of rivers in Britain. *Aquatic Conservation: Marine and Freshwater Ecosystems*, **8**, 597–616.

Boon, P. J., Holmes, N. T. H., Maitland, P. S. & Fozzard, I. R. (2002). Developing a new version of SERCON (System for Evaluating Rivers for Conservation). *Aquatic Conservation: Marine and Freshwater Ecosystems*, **12**, 439–55.

Bovee, K. D. (1982). A guide to stream habitat analysis using the instream flow incremental methodology. *United States Fish and Wildlife Service Biological Services Program Information Paper*, **12**, 1–248.

Briers, R. A. & Biggs, J. (2003). Indicator taxa for the conservation of pond invertebrate diversity. *Aquatic Conservation: Marine and Freshwater Ecosystems*, **13**, 323–30.

Cairns, J. (2000). Setting ecological restoration goals for technical feasibility and scientific validity. *Ecological Engineering*, **15**, 171–80.

Clark, M. J. (2002). Dealing with uncertainty: adaptive approaches to sustainable river management. *Aquatic Conservation Marine and Freshwater Ecosystems*, **12**: 347–63.

Clarke, S. J., Bruce-Burgess, L. & Wharton, G. (2003). Linking form and function: towards an eco-hydromorphic approach to sustainable river restoration. *Aquatic Conservation: Marine and Freshwater Ecosystems*, **13**, 439–50.

Cowx, I. G. & Welcomme, R. L. (eds.) (1998). *Rehabilitation of Rivers for Fish*. Oxford: Fishing News Books.

Davis, M. A. & Slobodkin, L. B. (2004). The science and values of restoration ecology. *Restoration Ecology*, **12**, 1–3.

Díaz-Pardo, E., Vazquez, G. & López-López, E. (1998). The phytoplankton community as a bioindicator of health conditions of Atezca Lake, Mexico. *Aquatic Ecosystem Health and Management*, **1**, 257–66.

Drake, D. C. & Naiman, R. J. (2000). An evaluation of restoration efforts in fishless lakes stocked with exotic trout. *Conservation Biology*, **14**, 1807–20.

Ebersole, J. L., Liss, W. J. & Frissell, C. A. (1997). Restoration of stream habitats in the western United States: restoration as re-expression of habitat capacity. *Environmental Management*, **21**, 1–14.

Forest Ecosystem Management Assessment Team (FEMAT) (1993). *Forest Ecosystem Management: An Ecological, Economic, and Social Assessment.* Report of the Forest Ecosystem Management Assessment Team. Washington, D.C: U.S. Government Printing Office.

Friberg, N., Kronvang, B., Hansen, H. O. & Svendsen, L. M. (1998). Long-term, habitat-specific response of a macroinvertebrate community to river restoration. *Aquatic Conservation: Marine and Freshwater Ecosystems*, **8**, 87–99.

Frissell, C. A. (1997). Ecological principles. In *Watershed Restoration: Principles and Practices*, eds. J. E. Williams, M. P. Dombeck & C. A. Wood. Bethesda, MD: The American Fisheries Society, pp. 96–115.

Frissell, C. A. & Bayles, D. (1996). Ecosystem management and the conservation of aquatic biodiversity and ecological integrity. *Journal of the American Water Resources Association*, **32**, 229–40.

Frissell, C. A. & Nawa, R. K. (1992). Incidence and causes of physical failure of artificial habitat structures in streams of Western Oregon and Washington. *North American Journal of Fisheries Management*, **12**, 182–97.

Frissell, C. A. & Ralph, S. C. (1998). Stream and watershed restoration. In *Ecology and Management of Streams and Rivers in the Pacific Northwest Coastal Ecoregion*, eds. R. J. Naiman & R. E. Bilby. New York: Springer-Verlag, pp. 599–624.

Frissell, C. A., Poff, N. L. & Jensen, M. E. (2001). Assessment of biotic patterns in freshwater ecosystems. In *A Guidebook for Integrated Ecological Assessments*, eds. P. Bourgeron, M. Jensen & G. Lessard. New York: Springer-Verlag, Chapter 27.

Frissell, C. A., Liss, W. J., Warren, C. E. & Hurley, M. D. (1986). A hierarchical framework for stream habitat classification: viewing streams in a watershed context. *Environmental Management*, **10**, 199–214.

Golet, G. H., Roberts, M. D., Luster, R. A. *et al.* (2006). Assessing societal impacts when planning restoration of large alluvial rivers: a case study of the Sacramento river project, California. *Environmental Management*, **37**, 862–79.

Hart, D. D., Johnson, T. E., Bushaw-Newton, K. L. *et al.* (2002). Dam removal: challenges and opportunities for ecological research and river restoration. *BioScience*, **52**, 669–81.

Heggenes, J. (1988). Physical habitat selection by brown trout (*Salmo trutta*) in riverine systems. *Nordic Journal of Freshwater Research*, **64**, 74–90.

Hilderbrand, R. H., Watts, A. C. & Randle, A. M. (2005). The myths of restoration ecology. *Ecology and Society*, **10**, 19.

Hildrew, A. G. & Giller, P. S. (1994). Patchiness, species interactions and disturbance in the stream benthos. In *Aquatic Ecology*, eds. P. S. Giller, A. G. Hildrew & D. G. Raffaelli. Oxford: Blackwell Science Ltd., pp. 21–62.

Holmes, T. P., Bergstrom, J. C., Huszar, E., Kask, S. B. & Orr, F. (2004). Contingent valuation, net marginal benefits and the scale of riparian ecosystem restoration. *Ecological Economics*, **49**, 19–30.

Hosper, S. H. (1998). Stable states, buffers and switches: an ecosystem approach to the restoration and management of shallow lakes in the Netherlands. *Water Science and Technology*, **37**, 151–64.

House, R. L. (1996). An evaluation of stream restoration structures on a coastal Oregon stream, 1981–1993. *North American Journal of Fisheries Management*, **16**, 272–81.

Howells, G. (1990). *Acid Rain and Acid Waters*. New York: Ellis Horwood: London.

Hynes, H. B. N. (1960). *The Biology of Polluted Waters*. Liverpool: Liverpool University Press.

Hynes, H. B. N. (1970). *The Ecology of Running Waters*. Liverpool: Liverpool University Press.

Hynes, H. B. N. (1975). The stream and its valley. *Verhandlungen der Internationalen Vereinigung für theoretische und angewandte Limnologie*, **19**, 1–15.

Irvine, K. (2004). Classifying ecological status under the European Water Framework Directive: the need for monitoring to account for natural variability. *Aquatic Conservation: Marine and Freshwater Ecosystems*, **14**, 107–12.

Janse, J. H. & van Liere, L. (1995). PCLAKE: a modelling tool for the evaluation of lake restoration scenarios. *Water Science and Technology*, **31**, 371–4.

Jordan, W. R., Gilpin, M. E. & Aber, J. D. (eds.) (1987). *Restoration Ecology*. Cambridge: Cambridge University Press.

Karr, J. R. (1991). Biological integrity: a long-neglected aspect of water resource management. *Ecological Applications*, **1**, 66–84.

Karr, J. R. & Dudley, D. R. (1981). Ecological perspective on water quality goals. *Environmental Management*, **5**, 55–68.

Kemp, J. L., Harper, D. M. & Crosa, D. A. (2000). The habitat-scale ecohydraulics of rivers. *Ecological Engineering*, **16**, 17–29.

Klinge, M., Grimm, M. P. & Hosper, S. H. (1995). Eutrophication and ecological rehabilitation of Dutch lakes: presentation of a new conceptual framework. *Water Science and Technology*, **31**, 207–18.

Knapp, R. A., Matthews, K. R. & Sarnelle, O. (2001). Resistance and resilience of alpine lake fauna to fish introductions. *Ecological Monographs*, **71**, 401–21.

Kondolf, G. M. (1995). Five elements for effective evaluation of stream restoration. *Restoration Ecology*, **3**, 133–6.

Kondolf, G. M. (1998). Lessons learned from river restoration projects in California. *Aquatic Conservation: Marine and Freshwater Ecosystems*, **8**, 39–52.

Langford, T. E. (1983). *Electricity Generation and the Ecology of Natural Waters*. Liverpool: Liverpool University Press.

Langford, T. E. (1990). *Ecological Effects of Thermal Discharges*. London & New York: Elsevier Applied Science.

Langford, T. E. (2006). Woody debris, restoration and fish populations. In *Fisheries on the Edge*, ed. I. J. Winfield. Proceedings of the Institute of Fisheries Management Conference 2005, Salford. Nottingham, UK: Institute of Fisheries Management.

Langford, T. E., Somes, J. R. & Bowles, F. (2000). Effects of physical restructuring on channels on the flora and fauna of three Wessex rivers. Pennington, Lymington, UK: Pisces Conservation Ltd.

Langford, T. E., Ferguson, A. G. P., Howard, S. R., Easton, K. & Shaw, P. J. (in press). Predicting recovery of ecological status in rivers using long-term data sets. In *Assessing the Ecological Status of Rivers, Lakes and Transitional Waters*, ed. I. G. Cowx. Proceedings of the HIFI Conference, Hull, September 2005.

Larsen, P. (1996). Restoration of river corridors: German experiences. In *River Restoration*, eds. G. Petts & P. Calow. Oxford: Blackwell Science. pp. 124–43.

Loomis, J., Kent, P., Strange, L., Fausch, K. & Covich, A. (2000). Measuring the total economic value of restoring ecosystem services in an impaired river basin: results from a contingent valuation survey. *Ecological Economics*, **33**, 103–17.

Maddock, I. (1999). The importance of physical habitat assessment for evaluating river health. *Freshwater Biology*, **41**, 373–91.

Madsen, B. L., Boon, P. J., Lake, P. S. *et al.* (2005). Ecological principles and stream restoration. *Verhandlungen der Internationalen Vereinigung für theoretische und angewandte Limnologie*, **29**, 2045–50.

Milner, A. M. (1996). System recovery. In *River Restoration*, eds. G. Petts & P. Calow. Oxford: Blackwell Science, pp. 205–26.

Milner, N. J., Hemsworth, R. J. & Jones, B. E. (1985). From THESIS Habitat evaluation as a fisheries management tool. *Journal of Fish Biology*, **27** (Suppl. A), 85–108.

Moerke, A. H. & Lamberti, G. A. (2004). Restoring stream ecosystems: lessons from a Midwestern state. *Restoration Ecology*, **12**, 327–4.

Moore, J. W., Schindler, D. E., Scheuerell, M. D., Smith, D. & Frodge, J. (2003). Lake eutrophication at the urban fringe, Seattle Region, USA. *AMBIO: A Journal of the Human Environment*, **32**, 13–18.

Moss, B. (1998). *Ecology of Fresh Waters*, 3rd edn. Oxford: Blackwell Science.

Moss, B., Stephen, D., Alvarez, C. *et al.* (2003). The determination of ecological status in shallow lakes – a tested system (ECOFRAME) for implementation of the European Water Framework Directive. *Aquatic Conservation: Marine and Freshwater Ecosystems*, **13**, 507–49.

Ormerod, S. J. (2004). A golden age of river restoration science? *Aquatic Conservation: Marine and Freshwater Ecosystems*, **14**, 543–9.

Palmer, M. A., Ambrose, R. F. & Poff, N. L. (1997). Ecological theory and community restoration ecology. *Restoration Ecology*, **5**, 291–300.

Palmer, M. A., Bernhardt, E. S., Allan, J. D. *et al.* (2005). Standards for ecologically successful river restoration. *Journal of Applied Ecology*, **42**, 208–17.

Pearce, F. (2006). *When The Rivers Run Dry*. Boston, MA: Beacon Press.

Petts, G. & Calow, P. (eds.) (1996). *River Restoration*. Oxford: Blackwell Science.

Prato, T. (2003). Multiple-attribute evaluation of ecosystem management for the Missouri River system. *Ecological Economics*, **45**, 297–309.

Ralph, S. C. & Poole, G. C. (2003). Putting monitoring first: designing accountable ecosystem restoration and management plans. In *Restoration of Puget Sound Rivers*, eds. D. Montgomery, S. Bolton & D. Booth. Seattle: University of Washington Press.

Riechert, P., Borsuk, M. Hostmann, M. *et al.* (2007). Concepts of decision support for river rehabilitation. *Environmental Modelling & Software*, **22**, 188–201.

River Restoration Centre (RRC) (1999). *Manual of River Restoration Techniques*. Silsoe, UK.

River Restoration Centre (RRC) (2002). *Manual of River Restoration Techniques*, 2nd edn. Silsoe, UK.

Roni, P. (ed.) (2005). *Monitoring Stream and Watershed Restoration*. Seattle, USA: Northwest Fisheries Science Center.

Sedell, J. R., Reeves, G. H., Hauer, F. R., Stanford, J. A. & Hawkins, C. P. (1990). Role of refugia in recovery from disturbances: modern fragmented and disconnected river systems. *Environmental Management*, **14**, 711–24.

Sethi, S. A., Selle, A. R., Doyle, M. W., Stanley, E. H. & Kitchel, H. E. (2004). Response of unionid mussels to dam removal in Koshkonong Creek, Wisconsin (USA). *Hydrobiologia*, **525**, 157–65.

Skinner, K. S. & Bruce-Burgess, L. (2005). Strategic and project level river restoration protocols – key components for meeting the requirements of the Water Framework Directive (WFD). *Water Environment*, **19**, 135–42.

Society for Ecological Restoration (2002) (Science & Policy Working Group). *The SER Primer on Ecological Restoration* (www.ser.org/).

Stanford, J. A., Ward, J. V., Liss, W. J. *et al.* (1996). A general protocol for restoration of regulated rivers. *Regulated Rivers: Research and Management*, **12**, 391–413.

Tockner, K., Schiemer, F. & Ward, J. V. (1998). Conservation by restoration: the management concept for a river-floodplain system on the Danube River in Austria. *Aquatic Conservation: Marine and Freshwater Ecosystems*, **8**, 71–86.

Usher, M. B. (ed.) (1986). *Wildlife Conservation Evaluation*. London: Chapman & Hall.

van Duin, E. H. S., Frinking. L. J., van Schaik, F. H. & Boers, P. C. M. (1998). First results of the restoration of Lake Geerplas. *Water Science and Technology*, **37**, 185–92.

Van Nes, E. H., Lammens, E. H. R. R. & Scheffer, M. (2002). PISCATOR, an individual-based model to analyze the dynamics of lake fish communities. *Ecological Modelling*, **152**, 261–78.

Vannote, R. L., Minshall, G. W., Cummins, K. W., Sedell, J. R. & Cushing C. E. (1980) The river continuum concept. *Canadian Journal of Fisheries and Aquatic Sciences*, **37**, 370–77.

Ward, J. V., Tockner, K., Uehlinger, U. & Malard, F. (2001). Understanding natural patterns and processes in river corridors as the basis for effective river restoration. *Regulated Rivers: Research & Management*, **17**, 311–23.

Welch, E. B. & Cooke, G. D. (1987). Lakes. In *Restoration Ecology*, eds. W. R. Jordan, M. E. Gilpin & J. D. Aber. Cambridge: Cambridge University Press, pp. 109–29.

Winterhalder, K., Clewell, A. F. & Aronson, J. (2004). Values and science in ecological restoration – a response to Davis and Slobodkin. *Restoration Ecology*, **12**, 4–7.

Woolsey, S., Capelli, F., Gonser, T. *et al.* (2007). A strategy to assess river restoration success. *Freshwater Biology*, **52**, 752–69.

Wright, J. F., Sutcliffe, D. W. & Furse, M. T. (eds.) (2000). *Assessing the Biological Quality of Fresh Waters. RIVPACS and Other Techniques*. Ambleside: Freshwater Biological Association.

Xu, F.-L., Tao, S., Dawson, R. W., Li, P.-G. & Cao, J. (2001). Lake ecosystem health assessment: indicators and methods. *Water Research*, **35**, 3157–67.

7 · *Methods for assessing the conservation value of rivers*

PHILIP J. BOON AND MARY FREEMAN

Introduction

Chapter 2 of this book sets out the context within which nature conservation value has been assessed for rivers in the UK and in the USA. Although the story is far from clear, there are two fundamental points that emerge: first, that the approach to river evaluation cannot be divorced from the purpose of the exercise; and second, that the methods used at present in both countries do not always mirror the views of those actively engaged in river ecology, management and conservation.

With respect to the first, Chapters 4, 5 and 6 (respectively) discuss the approaches used to assess river conservation values when the prime purpose is to select rivers for protection, to decide how to respond to the threat of environmental damage through catchment development or to plan a restoration programme where river habitats have become degraded. Regarding the second, the results from the snapshot survey of scientists, conservationists and resource managers (see Chapter 2) show that the retention of 'natural' features and processes should be the most important goal of river conservation.

To what extent are the assessment protocols available in the UK and in the USA able to take account of these requirements and aspirations? In both countries, methods of evaluation have developed in response to a number of drivers, some of which are legislative (e.g. the requirement to select rivers as Special Areas of Conservation under the EC Habitats Directive, or river segments under the US Wild and Scenic Rivers Act), whereas others are management-oriented (e.g. assessing the potential for intervention to restore riverine biodiversity). Yet the approaches to evaluation have often been piecemeal, lacking any overall strategy or structure. This chapter describes some of the methods used in the UK and in the USA for assessing river conservation value, and attempts to draw lessons for future application in both countries.

Assessing the Conservation Value of Fresh Waters, ed. Philip J. Boon and Catherine M. Pringle. Published by Cambridge University Press.

Methods used in international approaches

The past decade has seen a marked increase in the role of legislation in nature conservation, especially at an international level through a range of conventions, agreements and directives. Some of these are at a global level, and in principle affect the USA and the UK equally. Others have come from the European Union (EU), and influence much of the conservation effort of the UK and other EU Member States. None of these legislative mechanisms is directed specifically at river habitats and biota, but all, to some degree or other, contain measures that are relevant to them. (For background information see Chapter 2.)

Global

Ramsar Convention

Very few rivers have been designated as Ramsar sites, either for their habitat features or for their species. Instead, most sites are mosaics of 'true' wetland habitats, or they represent important transitions from open water to terrestrial habitats. A review of the summary descriptions for the Ramsar sites in five EU states (UK, France, Germany, Sweden and Greece) showed that more than 60% comprise three or more distinct wetland habitat types, such as wet grassland, wet heath, wet woodland, mudflats, seagrass beds and oxbow lakes. In the UK, the present suite of 168 Ramsar sites includes only a few stretches of rivers (Boon & Lee, 2005).

There are only 24 Ramsar sites in the USA (www.ramsar.org/sitelist.doc). As for the UK, these primarily comprise freshwater and estuarine wetlands protected for their bird life. Although some Ramsar sites include rivers within their boundaries (e.g. Horicon Marsh, Wisconsin, fed by the Rock River; Cypress Creek Wetlands, Illinois, fed by the Cache River), only two – Everglades National Park and White River National Wildlife Refuge – incorporate rivers in their own right as an integral part of the site's conservation importance.

Convention on Biological Diversity

In response to the Convention on Biological Diversity (CBD) the UK produced a Biodiversity Action Plan setting out a list of 37 'broad habitat types'. Within these, more specific work can be carried out on 'priority habitats' and 'priority species' (Boon & Lee, 2005). However, there is only one Habitat Action Plan (HAP) for running waters, aimed exclusively at chalk rivers, and the emphasis of HAPs is on action for maintenance or restoration rather than setting out detailed guidance for evaluating the rivers to be included in the plans. (At the time of writing (February 2008) a new broader 'Rivers' HAP was under development.) In the USA, there are no direct equivalents of CBD-derived HAPs for rivers. Instead, the Convention has helped to encourage a better understanding of aquatic systems through its Fisheries and Aquatic Resources Node of the Natural Biological Information Infrastructure (http://far.nbii.gov/), through published assessments of the state of the environment and through promoting protocols for freshwater monitoring.

IUCN Red Lists

Many approaches to conservation are centred on the assessment of rare or threatened species, either locally or regionally or sometimes at a national or international scale. Since the 1960s, the World Conservation Union (IUCN) has assessed the global conservation status of individual species (and lower taxonomic levels) according to the risk of extinction (www.iucn.org). The most recent IUCN Red List of Threatened Species (www.iucnredlist.org, 2004) contains 120 species for the USA in the habitat category 'Permanent Rivers/Streams/Creeks', including large numbers of mussels, fish and amphibians. In contrast, only two riverine species are listed for the UK, the moss *Thamnobryum angustifolium* and the otter *Lutra lutra*. Species on the Red List are classified within standard IUCN categories (e.g. 'endangered', 'vulnerable') using criteria such as 'population size' and 'extent of occurrence'. The usefulness of this general approach in river conservation evaluation is discussed later in the chapter.

European

Bern Convention

The Bern Convention was adopted and signed in Bern (Switzerland) in 1979, and came into force in 1982. The contracting parties comprise 39 Member States of the Council of Europe, as well as Burkina Faso, Monaco, Morocco, Senegal, Tunisia and the European Community. It aims to conserve wild flora and fauna and their natural habitats and to promote European cooperation in that field (www.coe.int/T/E/Cultural_Co-operation/Environment/Nature_and_biological_diversity/Nature_protection/_Summary.asp; www.jncc.gov.uk). Three of the four appendices to the Bern Convention list species that include some whose habitat is rivers (e.g. fish such as Atlantic salmon *Salmo salar* L., allis shad *Alosa*

alosa (L.) and river lamprey *Lampetra fluviatilis* (L.)). In the UK, implementation of the Convention is through the provisions of the Wildlife and Countryside Act 1981 and the EC Habitats Directive, both of which are described separately in this chapter.

EC Habitats and Birds Directives
Only one of the eight habitat types listed in Annex I of the Habitats Directive (H3260: 'Rivers with *Ranunculus*') occurs in the UK, limiting the overall proportion of riverine habitat designated as Special Areas of Conservation (SACs) and the degree to which SACs are able to represent areas for conservation under the UK Wildlife and Countryside Act (discussed below; also see Boon & Lee, 2005). There are 12 riverine species native to the UK that are listed on Annex II: one plant, three invertebrates, eight fish and one mammal. Of these, some are rare within the UK (e.g. freshwater pearl mussel *Margaritifera margaritifera*, spined loach *Cobitis taenia*) whereas others are not but are considered threatened at a European level (e.g. Atlantic salmon *S. salar*).

Some broad criteria for selecting riverine SACs are given in Annex III of the Habitats Directive. These include (for habitats) the area and degree of representativeness of the site covered by the natural habitat type, and (for species) the size and density of the population present on the site and the degree of isolation of the site's population in relation to the natural range of the species. Of course, these criteria have required further refinement in the UK to apply them to site selection. For example, for Atlantic salmon, sites were chosen as potential SACs on the basis of population size, genetic integrity of salmon populations, minimal environmental impacts (e.g from nutrient enrichment, organic pollution and channelization) and the degree to which the suite of rivers selected represents the species' geographical range and its ecological variants.

Methods used in the UK

Guidelines for selecting biological SSSIs

Until the advent of the EC Habitats Directive, the principal focus for river conservation evaluation in the UK was the selection of rivers as Sites of Special Scientific Interest (SSSIs) (Boon, 1995). Under the SSSI guidelines (Nature Conservancy Council, 1989), conservation evaluation is closely linked to a classification of River Community Types (RCTs) that was first drawn up in the early 1980s (Holmes, 1983). This was derived from aquatic macrophyte data on more than 200 rivers throughout Britain.

Table 7.1 *Classification of British rivers, based on their vegetation communities*

Group A	Lowland rivers with shallow gradients and rich geology
Type I	Lowland, low-gradient rivers
Type II	Lowland, clay-dominated rivers
Type III	Chalk rivers and other base-rich rivers with stable flows
Type IV	Impoverished lowland rivers
Group B	Meso-eutrophic rivers flowing predominantly over sandstone and hard limestone
Type V	Sandstone, mudstone and hard limestone rivers of England and Wales
Type VI	Sandstone, mudstone and hard limestone rivers of Scotland and northern England
Group C	Mesotrophic and oligo-mesotrophic rivers
Type VII	Mesotrophic rivers dominated by gravels, pebbles and cobbles
Type VIII	Oligo-mesotrophic rivers
Group D	Acid and nutrient-poor rivers
Type IX	Oligotrophic low-altitude rivers
Type X	Ultra-oligotrophic rivers

Source: Holmes *et al.* (1998).

The river classification scheme, revised using an expanded dataset (Holmes *et al.*, 1998), covers all of the main river types and divides British rivers into four groups (A–D), 10 RCTs (I–X) and 38 sub-types (AIa–DXe). Table 7.1 summarizes the classification to Group and RCT level. Geology is clearly a major factor differentiating the four groups, with altitude, size, flow, substrate and trophic status as important determinants for additional characterization.

The SSSI guidelines recommend that river SSSIs should be selected with the object of developing a national series to safeguard the best examples of the main types of river. In practice, however, most river SSSIs in Britain have been notified to underpin their designation as SACs, and the original intention expressed in the guidelines has not been fulfilled. The SSSI guidelines for rivers, whilst emphasizing the value of comprehensive biological surveys of plants, invertebrates, fish, birds and mammals, acknowledge that many sites will be selected based on the importance of their macrophyte communities alone, as these have been the principal focus of much of the river conservation survey work carried out in the UK over the past 25 years (Boon, 1995). There is a separate chapter in the guidelines on invertebrates, but this is mainly concerned with rare species, or with more specialized invertebrate assemblages such as those inhabiting river shingle beds. The guidelines for selecting SSSIs for freshwater fish have rarely been implemented, and they are generally considered insufficiently flexible to enable effective conservation of freshwater fish communities in Britain.

Community Conservation Index

The Community Conservation Index (CCI) was developed by Chadd and Extence (2004) to assess the conservation value of invertebrate communities in rivers and other freshwater habitats in Britain. It is one of the few examples in the UK of a conservation evaluation tool that uses a scoring system to provide a degree of consistency, although it focuses exclusively on one specific biological component of conservation value.

The CCI is based on two criteria – rarity and taxon richness. Each species recorded from a specific site or area is assigned a Conservation Score (CS) from 1–10, depending on its classification of rarity (Table 7.2). The sum of the CSs is calculated and then divided by the number of contributing species to give a mean score for the conservation value. This

Table 7.2 *Conservation Scores (CSs) for freshwater invertebrate species in Great Britain*

CS	Definition
10	RDB1 (Endangered)
9	RDB2 (Vulnerable)
8	RDB3 (Rare)
7	Notable (but not RDB status)
6	Regionally Notable
5	Local
4	Occasional (species not in categories 10–5, which occur in up to 10% of all samples from similar habitats)
3	Frequent (species not in categories 10–5, which occur in >10–25% of all samples from similar habitats)
2	Common (species not in categories 10–5, which occur in >25–50% of all samples from similar habitats)
1	Very Common (species not in categories 10–5, which occur in >50–100% of all samples from similar habitats)

Source: From Chadd & Extence (2004).

figure is multiplied by a Community Score (CoS), based on the Biological Monitoring Working Party Score (BMWP: Chesters, 1980) to introduce an element of species richness into the final CCI.

Methods used in the USA

The US government has promoted river conservation, either directly or indirectly, through at least three major legislative acts over the last 40 years. The Wild and Scenic Rivers Act (1968), the Clean Water Act (CWA) (1972) and the Endangered Species Act (ESA) (1973) have each supported some level of river assessment for conservation values, although based on different criteria (Table 7.3).

Assessing rivers for wild and scenic values

The Wild and Scenic Rivers Act (discussed in Chapter 2) facilitated a broadly based evaluation of the conservation value of free-flowing river segments across the USA. At the direction of Congress, the National Park Service undertook a nationwide inventory of river sections potentially eligible for inclusion in the wild and scenic rivers system. The National Rivers Inventory (NRI) identified more than 3400 undammed river segments in the USA with at least one scenic, recreational, cultural or natural feature of more than local or regional significance (www.nps.gov/ncrc/programs/rtca/nri). Value for biodiversity conservation was not necessary for inclusion in the NRI, nor was significant extent; some stream reaches of less than 20 km were listed, often reflecting isolated free-flowing segments in otherwise dammed or developed systems. Benke (1990) analysed a subset of NRI-listed streams (1524 segments that were >40 km in length) and noted that these constituted <2% of the total stream length in the USA, excluding Alaska and Hawaii.

Individual states also have wild and scenic river programmes that generally mirror the national programme in the intent to protect those streams and rivers of highest quality for the benefit of present and future generations. For example, the state of Ohio lists 20 stream segments in 11 river systems under its Scenic Rivers Act (www.ohiodnr.com/dnap/sr/). To be listed under Ohio's Act a stream must meet criteria establishing wild, scenic or recreational value; the criteria may include length, naturalness, water quality or biological characteristics. Listing requires the support of local governments and the public. This generally is true for listing as wild and scenic at either a state or federal level; meeting one or more criteria

Table 7.3 *Comparison of scale and features used to assess US streams and rivers for conservation value under differing legislative and assessment initiatives*

Initiative	Programmes or examples	Spatial scale	Features assessed for conservation value
Wild and Scenic River Act	National River Inventory, State Wild and Scenic rivers	Individual streams or stream segments	Recreational, scenic, geological, historical, or cultural values Fish and wildlife populations
Clean Water Act	Outstanding Resource Waters	Individual streams or stream segments	Water quality
Endangered Species Act	Critical Habitat designation	Individual streams or stream segments	Primary constituent elements, including: foraging, spawning, rearing and refuge habitat (e.g. appropriate flow regime, bed sediments, geomorphic structure); food or prey resources; water quality; host species (e.g. for mussels); lack of competing or predatory non-native species Level of threat or management needs
Status and trends assessments	State assessment and monitoring; federal programmes (e.g. NAWQA, EMAP-SW)	Stream reaches	Water chemistry, contaminants Physical habitat Biotic assemblages (periphyton, macroinvertebrates, fish) Sediment toxicity Community metabolism Fish tissue contaminants
Biodiversity assessments	North American freshwater ecoregions (Abell *et al.*, 2000)	Basins	Aquatic species richness and endemism (fish, mussels, reptiles and amphibians, crayfish) Occurrence of rare phenomena and habitats Human modification (land cover, water quality and hydrologic alteration; habitat loss and fragmentation; introduced species; species exploitation)
	Regional assessments (Smith *et al.*, 2002, Sowa *et al.*, 2003)	Sub-basins, stream networks	Rare or imperilled aquatic species Aquatic habitat representativeness Human disturbance

is a prerequisite but insufficient without strong public advocacy and government support for the listing.

Assessments based on outstanding water quality and biological integrity

The CWA (outlined in Chapter 2) and subsequent rulemaking provides a mechanism, the 'Outstanding Waters' designation, for identifying highest quality streams and rivers, again based on perceived aesthetic, environmental and ecological value to the public. States are not actually required to designate any waters as 'Outstanding', but if they do the federal Environmental Protection Agency (EPA) may also recognize those bodies as 'Outstanding National Resource Waters', which would require the strictest level of water quality protection. States have varied in their application of the 'Outstanding Waters' designation, but generally apply criteria that reflect the notion embodied in NRI recognition that the water body has exceptional ecological or environmental qualities. Citizens and advocacy groups may nominate waters for 'Outstanding' designation in some states; however, states generally lack a systematic process with specific criteria for designating streams and rivers as 'Outstanding Waters' under the CWA.

The CWA has also fostered the development of monitoring and assessment methods, and publicly available datasets, potentially useful in assessing conservation value. Under the Act each state is required to prepare and submit to the EPA a biennial report on the water quality conditions of its navigable waters. State monitoring programmes vary widely; however, the EPA encourages the use of physical, chemical and biological indicators to assess the attainment of water quality standards. The emphasis of monitoring and assessment is on tracking trends in water quality, identifying impaired waters and the causes of impairment, and evaluating success of management activities. Yet the data and methods developed for assessments also provide a potential basis for defining the conservation value of streams and rivers with respect to biological and habitat integrity (discussed further under 'Progress in the USA').

Assessing critical habitat for imperilled species

The process of designating 'critical habitat' for species listed as threatened or endangered under the federal ESA (see Chapter 2) provides a basis for assessing conservation value of streams and rivers relative to their importance to sustaining imperilled species. At present, critical habitat has been designated for 59 of 114 freshwater fish and 18 of 70 freshwater mussel species listed as endangered or threatened in the USA. Assessment criteria are tailored for the targeted species, but generally include occurrence of 'primary constituent elements' – those physical and biological features required by the species over its life cycle (Table 7.3; US Fish and Wildlife Service (USFWS), 2004). Critical habitat may include areas that are not at present occupied by the species if those areas have appropriate habitats and if the occupied range is considered inadequate for long-term species conservation. Moreover, because a river or stream segment can only be designated as critical habitat if existing management is insufficient to protect that habitat (USFWS, 2004), the process also implicitly identifies areas that are in greater need of conservation action than others.

Towards a more consistent approach

Progress in the UK

In the UK, the evaluation of rivers for conservation has been hampered by the lack of consistent and repeatable methods. Most work has been in response to the need to select rivers for protection, using the broad principles of the SSSI guidelines, or the criteria limited to the few riverine habitats and species listed in the Habitats Directive. Alternatively, statements concerning the specific conservation value of particular rivers or river stretches have been made from time to time, but without any overall framework to establish context and allow comparison between one river and another.

The mid-1990s saw the coincident development of two river evaluation systems, both of which are now closely linked in providing a broad-based, repeatable method for assessing the value of river habitats and biota. The System for Evaluating Rivers for Conservation (SERCON) was an inter-agency initiative that commenced in the early 1990s, while the development of River Habitat Survey (RHS) began soon afterwards by the Environment Agency (EA) of England and Wales.

River Habitat Survey

The background, structure and function of RHS were described in detail by Raven *et al.* (1997). RHS was conceived as a nationally applicable system for measuring, classifying and reporting on the physical structure of rivers. It was not devised specifically as a system for conservation assessment, but its value in providing river conservation with a tool for describing the status of river habitats soon became obvious. Indeed, its approach was based on establishing a national inventory of features considered to be important for wildlife (Harper *et al.*, 1995), and the rationale for its development included the need to monitor SACs under the Habitats Directive, and to monitor physical structure as a surrogate for biodiversity as part of the UK Biodiversity Action Plan (Raven *et al.*, 1997). RHS is based on a standard field survey method, undertaken by surveyors who have received training in its use, and who have passed an accreditation test. Whilst surveyors do not need to be geomorphological or botanical specialists, they do require an ability to recognize vegetation types and understand broad geomorphological principles. The field method has been revised recently (Environment Agency, 2003) and is expected to remain unchanged for at least 5 years.

RHS is carried out on standard 500 m lengths of river channel, where observations are made at ten locations (spot-checks) spaced evenly along the survey site. Those features recorded include channel substrate, channel and bank modifications and bankface vegetation structure. In addition, general information on the character of the RHS site is made using a 'sweep-up' checklist, noting features such as adjacent land use and invasive plant species. This represents an assessment of the extent of features over the whole 500 m length, and includes features that do not occur at spot-checks. Field data are recorded on a four-page form, and later validated by the EA before adding to the national database. At present, the database contains records from around 17 000 sites in the UK, including *c.* 4600 baseline survey sites comprising a geographically representative cross section of rivers and streams selected using a stratified random sampling approach.

As well as compiling feature lists, two habitat indices can be derived from RHS. Habitat Modification Class (HMC) is an index of 1–5, where 1 represents the lowest degree of modification. The index reflects the extent and type of modifications affecting the natural features and processes of rivers (e.g. hard reinforcement of channel and banks, resectioning of bed or banks) together with an assessment of the resilience of such modifications. Habitat Quality Assessment (HQA, also scored from 1–5) combines scores for diversity of features such as substrates, flow types and channel vegetation structure with those for the presence of unusual or particularly valued conservation features of rivers and floodplains.

SERCON: System for Evaluating Rivers for Conservation

There were two principal motives for developing a new technique for river conservation evaluation in the UK: a perceived need to increase the breadth, rigour and repeatability of evaluations, and a desire to shift the emphasis in river conservation from trying only to protect the best to managing, improving and restoring river resources across the full spectrum of conservation value.

Work on SERCON began in the early 1990s, led by Scottish Natural Heritage in partnership with the statutory conservation and environment agencies in the UK. Detailed descriptions of the development, structure and function of SERCON have been published elsewhere (Boon *et al.*, 1997, 2002); this chapter briefly highlights its main features and describes the current status of the system.

SERCON is a tool for assessing a wide range of the biological and habitat attributes of rivers. Each evaluation is restricted to a river reach known as an 'ECS' – an abbreviation for 'Evaluated Corridor Section'. An ECS is essentially a natural river reach, usually between 10 and 30 km, with uniform gross physical characteristics of geology, slope, size, etc.

SERCON comprises a combination of scored and unscored elements. Scoring is incorporated to ensure repeatability, and to enable comparisons between ECSs. Those attributes that are scored relate to a suite of traditional conservation criteria such as naturalness, representativeness, species richness and rarity (Ratcliffe, 1977; Boon *et al.*, 1997; see also Chapter 2), and to a range of human impacts (Table 7.4). Those that are unscored are included either as background data (e.g. stream order, land use) to set the context in which evaluations are made, or so that important features can be highlighted which cannot be evaluated using the scored criteria. The outputs from both parts of SERCON are essential to a balanced interpretation of the conservation value of an ECS.

SERCON is designed so that individual river attributes contribute towards an understanding of each conservation criterion (Table 7.4). Moreover, a single biological dataset, interpreted in different ways, may be used to contribute to different conservation criteria. For example, data on freshwater fish are used in the assessment of three criteria: Naturalness, Rarity and Species Richness. The numerical output from SERCON is a set of attribute scores (on a scale of 0–5), and a suite of conservation and impact indices (on a scale of 0–100) derived by combining and weighting individual scores. An unscored category (Additional Features of Importance) allows the user to draw attention to unique or unusual features of the river.

Table 7.4 *List of conservation attributes evaluated in SERCON 2, grouped within the criteria of physical diversity, naturalness, representativeness, rarity, species richness, and special features. The 11 riverine impacts evaluated are also listed.*

Physical Diversity
PDY 1: Channel Substrates
PDY 2: Flow-types and Habitat Features
PDY 3: Structure of Aquatic Vegetation

Naturalness
NA 1: Planform and River Profile
NA 2: Extent of Channel and Bank Engineering
NA 3: Channel and Bank Features
  NA 3a: Habitat Quality Assessment
  NA 3b: Habitat Modification Class
NA 4: Flow Regime
NA 5: Plant Assemblages on the Banks
NA 6: Riparian Zone
NA 7: Aquatic and Marginal Macrophytes
NA 8: Aquatic Macroinvertebrates
NA 9: Fish
NA 10: Breeding Birds

Representativeness
RE 1: Aquatic Macrophytes

Rarity
RA 1: EC Habitats Directive Species (+ rare in UK)
RA 2: Scheduled Species
RA 3: Red List Species
RA 4: EC Habitats Directive Species (but not rare in UK)
RA 5: Other Species Under Threat
RA 6: Macrophyte Species Uncommon in England, Wales, Scotland or Northern Ireland
RA 7: Breeding Bird Species Uncommon in England, Wales, Scotland or Northern Ireland

Species Richness
SR 1: Aquatic and Marginal Macrophytes
SR 2: Aquatic Macroinvertebrates
SR 3: Fish
SR 4: Breeding Birds

Special Features
SF 1: Complexity and Character of Riparian Zone
SF 2: Corridor Water-dependent Habitats
SF 3: Marginal Habitats for Invertebrates
SF 4: Wintering Birds on Floodplain
SF 5: Other Vertebrates

Table 7.4 (*cont.*)

Impacts	
IM 1:	Acidification
IM 2:	Urban, Industrial and Agricultural Inputs
IM 3:	Sewage Effluent
IM 4:	Groundwater Abstraction
IM 5:	Surface Water Abstraction
IM 6:	Inter-river and Inter-basin Transfers
IM 7:	Channelization
IM 8:	Management for Flood Defence
IM 9:	Artificial Structures
IM 10:	Recreational Pressures
IM 11:	Introduced Species

The first version of SERCON was completed in 1996, both as a printed manual and as a software program, and was tested on a wide range of rivers in Britain. A comprehensive review of SERCON was completed in 1999 and led to the development of an improved version. New features include the derivation of all information on physical features from RHS. The review and revision of SERCON highlighted one particular dilemma. Although it is generally accepted that nature conservation in rivers needs to be broadly based, assembling this range of information is often time-consuming and may be prohibitively expensive (Boon, 2000; Boon *et al.*, 2002). There may also be other reasons why a full SERCON evaluation may not be feasible, or even necessary. For example, the only reason for assessment might be to determine the importance of specific animal or plant groups, or an environmental impact assessment is required only for a limited number of features. SERCON 2 therefore includes a series of 'SERCON Applications' – individual modules (e.g. 'River Rehabilitation', 'Site Monitoring') that comprise groups of attributes selected by the user.

Progress in the USA

Conservation 'value' has been defined in the USA, through the legislation discussed above, to include protection of native biodiversity and of natural systems that are relatively unaltered by human activities – goals explicit in initiatives such as Ramsar, the EC Habitats Directive and SERCON. Methods to assess these values in lotic systems have developed along two broad directions. These approaches reflect differing emphases on measuring site-specific physical and biological integrity (as in status and trends assessments, Table 7.3) as opposed to identifying sites or catchments important to conserving biodiversity.

Assessing stream and river integrity

Federal and state agencies have designed and implemented a suite of assessment protocols to measure the physical and biological condition of streams and rivers. Initiatives such as the EPA's Environmental Monitoring and Assessment Program – Surface Waters (EMAP-SW), the US Geological Survey's National Water-Quality Assessment (NAWQA) Program, and assessments by state environmental agencies, utilize site-specific measurements of channel features and biota to assess physical and biological conditions relative to regional reference conditions. Assessments typically include physical habitat (e.g. channel dimensions, bed sediments, riparian vegetation) and one or more biotic groups (periphyton, macroinvertebrate and fish assemblages), measured using standardized protocols (Lazorchak *et al.*, 1998; Barbour *et al.*, 1999; Moulton *et al.*, 2002). These assessments extend water quality monitoring programmes so as to capture degradation not detectable by periodic water sampling (Karr, 1991). The general goal of state programmes is to detect stream impairment (i.e. with respect to CWA goals). Goals of the national programmes include assessing status and trends of freshwater ecological resources within regions (EMAP-SW; Lazorchak *et al.*, 1998) and across major river basins (NAWQA). The national programmes also seek to relate status to potential causal mechanisms such as point or non-point sources of pollutants, channel alteration or changes in stream flow patterns.

Assessments of stream integrity are targeted frequently to areas suspected or identified as degraded, or used to assess the results of mitigation efforts. Site-specific assessments typically are made in reaches measuring a few hundred metres (e.g. often 20–40 times the channel width). Assessments at locations distributed across catchments using a probabilistic sampling design have potential use in assessing the conservation value of stream systems based on standardized measures of physical and biological integrity. Catchments with consistently high scores for integrity might be valued more highly for conservation than catchments with evidence of widespread impairment. However, protocols for assessing streams relative to reference conditions do not generally assess the specific value of a site for conserving unique biodiversity, such as rare or imperilled species or distinct community types.

Moyle and Randall (1998) evaluated a catchment-level index of biotic integrity that specifically focused on catchment value for protecting regionally unique biodiversity in the Sierra Nevada, California. In this application, catchments were scored for up to six metrics based on the occurrence and abundance of native fish and frogs using narrative criteria (e.g. native fish 'absent or rare', present only in portions of their native range but 'still easy to find' or found throughout the catchment at 'presumed historic levels'). The authors recommended focusing conservation efforts on developing a system of protected areas with high catchment-level biotic integrity, and representative of the range of aquatic environments in the region.

Assessing value for freshwater biodiversity

Emphasis on protecting exceptional freshwater biodiversity in the USA has fostered approaches for assessing aquatic conservation value on the basis of species diversity, biological distinctiveness and level of threats or degradation (Table 7.3). Rabe and Savage (1979), for example, proposed evaluating sites to be protected as 'aquatic natural areas' using a numerical ranking system based on attributes such as the diversity of aquatic habitats, unusual water chemistry, high species diversity, and species considered rare, uncommon or of special concern. The importance of the highest-ranking sites for protection would be assessed according to ownership, size, accessibility, threats, and the extent to which similar habitat features are already represented in established natural areas. At a continental scale, the World Wildlife Fund – USA (Abell *et al.*, 2000) similarly assessed conservation priority levels for 76 North American freshwater ecoregions (basins or groups of basins having distinct biotic assemblages) on the basis of biological diversity and uniqueness, and level of degradation. Experts provided data to rank the biological distinctiveness of each ecoregion within eight major habitat types (reflecting latitudinal and climatic gradients), on the basis of four criteria: (1) species richness and number of endemics (for fish, mussels, crayfish, aquatic reptiles and amphibians); (2) presence of 'rare ecological or evolutionary phenomena' (e.g., large-scale fish migrations); (3) the occurrence of high species turnover along environmental gradients; and (4) the occurrence of globally rare habitats. Based on these four criteria, ecoregions were rated as 'globally outstanding', 'continentally outstanding', 'bioregionally outstanding', or 'nationally important' (the least biologically distinctive level). Conservation status, ranging from 'relatively intact' to 'critical', was assessed also for each ecoregion on the basis of scores for seven criteria reflecting human modification. The outcome was a ranking of North American freshwater ecoregions, grouped by habitat type, for conservation efforts.

Regional assessments of stream value for biodiversity have been based on occurrences of target species and habitats (Table 7.3). Ideally, targets are chosen to represent the spectrum of biodiversity in a system or across a landscape; however, target species are necessarily limited to taxa (typically aquatic vertebrates, molluscs and crayfish) for which data are sufficient to describe or model geographic distributions. Even for well-studied groups such as fish, the occurrence of undescribed, cryptic diversity (Burkhead & Jelks, 2000) points to the potential inadequacy of relying solely on species targets. Thus, targeting habitat types is intended as a 'coarse filter' to ensure broad representation of biodiversity not covered by a strategy focused on special elements such as rare species (Groves *et al.*, 2002; Noss, 2004).

Defining target habitats for stream communities has required methods for classifying systems that typically exist within continuous lotic networks. One approach has been to use biological data to distinguish community types, which may subsequently be linked to physical settings. For example, Angermeier and Winston (1999) used multivariate analyses of catch data to assess the diversity of stream fish communities within the state of Virginia; the result was about 90 distinct community types arrayed across differing drainages, physiographic regions, elevations and stream sizes. An alternative approach has involved directly classifying lotic habitats on the basis of hierarchically nested features that create faunal or physical discontinuities, such as basin divides and physiographic provinces, together with categorical analysis of variation in physical features such as geology, stream size, elevation, gradient and hydrology (Groves *et al.*, 2002; Smith *et al.*, 2002; Higgins *et al.*, 2005). For example, in an aquatic evaluation under the National Gap Analysis Program (Jennings, 2000), biologists targeted 'aquatic ecological systems' (stream networks distinguished by geologic, hydrologic and geomorphic features, Higgins *et al.*, 2005) within separate basins or sub-basins that supported differing aquatic assemblages (Sowa *et al.*, 2003). The conservation value of individual aquatic ecological systems was assessed based on predicted richness of target species (e.g. fish, mussels and crayfish considered to be of conservation concern), degree of human disturbance, percentage area in public ownership, and other considerations such as historic disturbances or likelihood of landowner cooperation (Sowa *et al.*, 2003).

Defining target habitats as physically distinctive portions of stream systems for conservation assessment assumes that the habitat classifications actually represent the scale at which ecological processes drive community variation. Conroy and Noon (1996), for example, have shown the potential mismatch in the scale of habitats identified through 'coarse-filter' approaches and the scales at which demographic processes shape population densities. However, methods for assessing conservation value explicitly with respect to ecological or demographic processes are not yet well developed. Winston and Angermeier (1995) have proposed assessing conservation value of stream sites on the basis of relative densities of targeted fish species, assuming that measured densities reflect importance of the habitat to source populations. Angermeier and Winston (1997) note that the 'index of centres of density', while potentially valuable for assessing the conservation values of sites that lack regionally rare or imperilled species, could undervalue geographically peripheral populations that exist at relatively low densities and yet contain important genetic variation. Ultimately, quantifying population viability in relation to natural and human impact variables will require data on population demography, including dispersal (Noss, 2004).

Discussion

This chapter has set out to examine how rivers are evaluated for nature conservation in the UK and in the USA. In common with the other chapters in this book, the subject has been restricted to evaluating rivers in terms of their physical and chemical features, their biological communities, and their ecological processes. This is not intended to underestimate the wider value of rivers to society: whether as economic, aesthetic or recreational resources (Boon & Howell, 1997). Indeed, in the USA, the Wild and Scenic Rivers Act specifically draws its net wide to capture rivers considered important for landscape and recreation, as well as for ecology.

Viewed from the narrower perspective of habitats and species assessment, methods developed in the USA and the UK share a common emphasis on classifying running water habitats in order to represent the range of aquatic biodiversity and thereby to protect 'biological distinctiveness'. There are both similarities and differences, however, in the way that biological distinctiveness is perceived and evaluated, and in the effectiveness with which rivers are selected for conservation.

One of the common aims of river conservation strategies in both the USA and the UK is to ensure the inclusion of representative 'river types' on any list of protected areas. By definition, this strategy is based on systems of classification (or 'typology') so that the best examples of each type can be selected. In the USA, high species richness and, especially, community distinctiveness even among nearby drainages, results in many more conservation targets than in the UK. For example, Holmes *et al.* (1998) identified 38 River Community sub-types in the UK based on their aquatic macrophyte communities. This compares with 115 aquatic habitat targets in a single US river basin (Smith *et al.*, 2002) or 90 potential fish community targets in one state (Angermeier & Winston, 1999). Perhaps the strength of the US approach is its ability to identify and capture biological diversity at a finer scale. On the other hand, its weakness lies in the requirement for a huge amount of supporting data (e.g. on species distributions) and the need to make subjective decisions in the classification process about which physical variables (e.g. stream size, gradient, geology) are most relevant to variation in biological communities, and at what scale.

One of the more significant, and positive, trends in both the UK and the USA over the past 20–30 years is a widening of the canvas on which river conservation assessments are made. This is seen in three distinctly different ways. First, it is becoming more generally accepted that river conservation cannot work if rivers are seen merely as channels transporting water: they must be viewed as river 'corridors' comprising channel, banks and riparian zone (Ormerod, 1999). Where there are floodplains, maintaining their connectivity with the river channel is essential; and the ultimate goal of integrated catchment management is beginning (at least to some extent) to climb the political agenda in both countries.

Second, as the goals of river management and river conservation have begun to move closer (see Chapter 3; Boon, 2000), so the way that rivers are surveyed and monitored has begun to expand. Historically, river management in both countries has largely centred on water quality, with rafts of legislation to deal with the more obvious forms of industrial, domestic and urban pollution. Both in the UK and in the USA, the last 20–30 years has seen an increasing and welcome emphasis on biological monitoring, usually involving macroinvertebrates, rather than the routine laboratory analysis of water samples for a limited range of chemical variables (Hellawell, 1997; Wright *et al.*, 1997; Karr *et al.*, 2000).

Third, and of greatest significance for river conservation, there is a degree of consensus that water quality and invertebrate communities are only two aspects of a far more comprehensive list of attributes that should be used to assess river conservation value. Despite its obvious limitations, the 'snapshot' survey of freshwater scientists, conservationists and resource managers (see Chapter 2) indicates that there are today broad areas of agreement (both across the range of sectors and between the UK and the USA) on those aspects of rivers deemed important for conservation. There is overwhelming support for the concept of 'naturalness' as the most important criterion in river conservation, followed closely by the need to conserve rare and threatened species. Other conservation criteria, such as 'representativeness' (or 'typicalness') and species richness are also considered valuable by many, although to a lesser degree.

The maintenance or restoration of naturalness as an aim of conservation is one that has been frequently debated (Angermeier, 1994; Hunter, 1996). To some extent, the arguments are as much about philosophy as they are about ecology, covering issues such as the place of human society in nature, and the intrinsic value that many conservationists place on natural as opposed to artificial features. Angermeier (2000) summarizes this debate in an interesting critique of the role of naturalness as a guiding concept in conservation. He argues that naturalness is a legitimate goal of conservation; that the objective definition of naturalness is both desirable and possible; and that naturalness is a continuous variable allowing degrees of naturalness to be identified. This view is entirely in accord with the methods (such as SERCON) described in this chapter for evaluating rivers, and is further supported by the views canvassed in our questionnaire.

These perceptions of value are, to a limited extent, beginning to work their way into practice. Instead of concentrating biological sampling in rivers on macroinvertebrates for pollution monitoring, or fish for fisheries management purposes, other biotic groups such as aquatic macrophytes are now receiving attention. Instead of seeing rivers as fixed in space and time, there is an increasing awareness of their dynamism; and that to conserve riverine biodiversity (in its broadest sense) requires an understanding of the geomorphological processes that shape river habitats, and the ecological processes that structure biological communities (Newson, 2002).

Yet, this still has far to go. Wider visions of what a river of high conservation value should look like are not reflected particularly well in environmental legislation or in general conservation practice, perhaps more especially in the USA. This is chiefly a matter of balance and focus. One area of particular concern is the extent to which conservation effort is frequently targeted at individual species – usually those that are threatened, those with public appeal, or those for which data are more available (such as fish, molluscs and crayfish in the USA) – rather than on communities or habitats. This approach is encouraged by legislation such as the ESA in the USA, and the schedules of the Wildlife and Countryside Act in the UK. Even those more recent international initiatives, such as the CBD aimed at promoting biological conservation in the broadest sense, have been worked out at a national and local level with a more limited focus on threatened habitats and biota.

For rivers, the danger is not only that biodiversity in general will not be adequately conserved, but also that the message promoted is an incomplete one. Where emphasis is placed on the rare and threatened, the commonplace is undervalued. Where the focus is on protected 'sites' for conservation, the 'wider countryside' of which those sites are an integral part may be neglected (Boon & Lee, 2005).

In the UK, impetus for further change has come with the implementation of the EC Water Framework Directive (WFD) (see Chapter 2; European Commission, 2000). The WFD covers all surface waters (rivers, lakes, estuaries, coastal waters) as well as groundwater. One of its primary aims is to prevent further deterioration of aquatic ecosystems and to protect and enhance their status. At the heart of the Directive is a requirement to produce river basin management plans, with the aim of achieving 'good surface water status' by 2015 (Boon & Lee, 2005). The WFD is the most comprehensive legislation ever enacted in Europe to address the integrity of freshwater ecosystems. Although it is not a 'conservation' directive, and the concept of 'ecological status' enshrined within it is not synonymous with 'nature conservation value', the Directive does have much to contribute to the field of aquatic conservation. In particular, it now provides a statutory basis for wider environmental assessment – in terms of spatial coverage (river basins), habitat attributes (morphology, hydrology, water quality, riparian zones) and biological communities (macrophytes, phytobenthos, phytoplankton, benthic invertebrates, fish) – and makes river basin management a mandatory requirement in fulfilling its objectives.

Of course, the challenge in river conservation is to find ways of moving from survey and inventory into evaluation, from evaluation into conservation planning, and from conservation planning into practical conservation management. Both in the US and in the UK there have been few examples of entire river systems (even small ones) protected for their conservation value. Instead, river reaches, or occasionally the whole of a main stem or tributary, have been selected as conservation units. Whereas policy and legislation may struggle with taking a more expansive approach, geomorphology and biology emphasize that conservation requires protecting the landscape-scale processes that generate and maintain ecological communities (Angermeier & Karr, 1994). In stream systems, processes such as the dispersal and migration of organisms along linear networks, and the water-mediated flow of materials and energy, make it almost impossible to protect any particular segment entirely separately from upstream and downstream areas, and from terrestrial influences (Pringle, 2001; Saunders *et al.*, 2002). The challenge for managing rivers in the new millennium is surely to bring policy and practice into line with a scientific understanding of how rivers work.

Acknowledgements

We thank Paul Angermeier for review comments that greatly improved the text.

References

Abell, R. A., Olson, D. M., Dinerstein, E. *et al.* (2000). *Freshwater Ecoregions of North America: A Conservation Assessment.* Washington, DC: Island Press.

Angermeier, P. L. (1994). Does biodiversity include artificial diversity? *Conservation Biology*, **8**, 600–2.

Angermeier, P. L. (2000). The natural imperative for biological conservation. *Conservation Biology*, **14**, 373–81.

Angermeier, P. L. & Karr, J. R. (1994). Biological integrity versus biological diversity as policy directives. *BioScience*, **44**, 690–8.

Angermeier, P. L. & Winston, M. R. (1997). Assessing conservation value of stream communities: a comparison of approaches based on centres of density and species richness. *Freshwater Biology*, **37**, 699–710.

Angermeier, P. L. & Winston, M. R. (1999). Characterizing fish community diversity across Virginia landscapes: prerequisite for conservation. *Ecological Applications*, **9**, 335–49.

Barbour, M. T., Gerritsen, J., Snyder, B. D. & Stribling, J. B. (1999). *Rapid Bioassessment Protocols for Use in Streams and Wadeable Rivers: Periphyton, Benthic Macroinvertebrates and Fish*, 2nd edn. EPA 841-B-99-002. Washington, DC: U. S. Environmental Protection Agency, Office of Water.

Benke, A. C. (1990). A perspective on America's vanishing streams. *Journal of the North American Benthological Society*, **9**, 77–88.

Boon, P. J. (1995). The relevance of ecology to the statutory protection of British rivers. In *The Ecological Basis for River Management*, eds. D. M. Harper & J. D. Ferguson. Chichester: John Wiley, pp. 239–50.

Boon, P. J. (2000). The development of integrated methods for assessing river conservation value. *Hydrobiologia*, **422/423**, 413–28.

Boon, P. J. & Howell, D. L. (eds.) (1997). *Freshwater Quality: Defining the Indefinable?* Edinburgh: The Stationery Office.

Boon, P. J. & Lee, A. S. L. (2005). Falling through the cracks: Are European directives and international conventions the panacea for freshwater nature conservation? *Freshwater Forum*, **24**, 24–37.

Boon, P. J., Holmes, N. T. H., Maitland, P. S. & Fozzard, I. R. (2002). Developing a new version of SERCON (System for Evaluating Rivers for Conservation). *Aquatic Conservation: Marine and Freshwater Ecosystems*, **12**, 439–55.

Boon, P. J., Holmes, N. T. H., Maitland, P. S., Rowell, T. A. & Davies, J. (1997). A system for evaluating rivers for conservation ('SERCON'): development, structure and function. In *Freshwater Quality: Defining the Indefinable?*, eds. P. J. Boon & D. L. Howell. Edinburgh: The Stationery Office, pp. 299–326.

Burkhead, N. M. & Jelks, H. L. (2000). Diversity, levels of imperilment, and cryptic fishes in the southeastern United States. In *Freshwater Ecoregions of North America: A Conservation Assessment*, eds. R. A. Abell, D. M. Olson, E. Dinerstein *et al.* Washington, DC: Island Press, pp. 30–2.

Chadd, R. & Extence, C. (2004). The conservation of freshwater macroinvertebrate populations: a community-based classification scheme. *Aquatic Conservation: Marine and Freshwater Ecosystems*, **14**, 597–624.

Chesters, R. K. (1980). *Biological Monitoring Working Party. The 1978 National Testing Exercise.* Technical Memorandum No. 19. Department of the Environment Water Data Unit, London.

Conroy, M. J. & Noon, B. R. (1996). Mapping of species richness for conservation of biological diversity: conceptual and methodological issues. *Ecological Applications*, **6**, 763–73.

Environment Agency (2003). *River Habitat Survey in Britain and Ireland. Field Survey Guidance Manual: 2003 Version.* Bristol: Environment Agency.

European Commission (2000). Directive 2000/60/EC, Establishing a framework for community action in the field of water policy. *Official Journal of the European Communities*, L **327**, 1–71.

Groves, C. R., Jensen, D. B., Valutis, L. L. *et al.* (2002). Planning for biodiversity conservation: putting conservation science into practice. *BioScience*, **52**, 499–512.

Harper, D. M., Smith, C., Barham, P. & Howell, R. (1995). The ecological basis for the management of the natural river environment. In *The Ecological Basis for River Management*, eds. D. M. Harper & A. J. D. Ferguson. Chichester: John Wiley, pp. 219–38.

Hellawell, J. M. (1997). The contribution of biological and chemical techniques to the assessment of water quality. In *Freshwater Quality: Defining the Indefinable?*, eds. P. J. Boon & D. L. Howell. Edinburgh: The Stationery Office, pp. 89–101.

Higgins, J. V., Bryer, M. T., Khoury, M. L. & Fitzhugh, T. W. (2005). A freshwater classification approach for biodiversity conservation planning. *Conservation Biology*, **19**, 432–45.

Holmes, N. T. H. (1983). *Typing British Rivers According to their Flora.* Focus on Nature Conservation 4. Peterborough: Nature Conservancy Council.

Holmes, N. T. H., Boon, P. J. & Rowell, T. A. (1998). A revised classification system for British rivers based on their aquatic plant communities. *Aquatic Conservation: Marine and Freshwater Ecosystems*, **8**, 555–78.

Hunter, M. L. (1996). Benchmarks for managing ecosystems: are human activities natural? *Conservation Biology*, **10**, 695–7.

Jennings, M. D. (2000). Gap analysis: concepts, methods, and recent results. *Landscape Ecology*, **15**, 5–20.

Karr, J. R. (1991). Biological integrity: a long-neglected aspect of water resource management. *Ecological Applications*, **1**, 66–84.

Karr, J. R., Allan, J. D. & Benke, A. C. (2000). River conservation in the United States and Canada. In *Global Perspectives on River Conservation: Science, Policy and Practice*, eds. P. J. Boon, B. R. Davies & G. E. Petts. Chichester: John Wiley, pp. 3–39.

Lazorchak, J. M., Klemm, D. J. & Peck, D. V. (eds.) (1998). *Environmental Monitoring and Assessment Program – Surface Waters. Field Operations and Methods for Measuring the Ecological Condition of Wadeable Streams*. EPA/620/R-94/004F. Washington, DC: U.S. Environmental Protection Agency.

Moulton, S. R., Kennen, J. G., Goldstein, R. M. & Hambrook, J. A. (2002). *Revised Protocols for Sampling Algal, Invertebrate, and Fish Communities as Part of the National Water-Quality Assessment Program*. U.S. Geological Survey Open-File Report 02-150, Reston, VA.

Moyle, P. B. & Randall, P. J. (1998). Evaluating the biotic integrity of watersheds in the Sierra Nevada, California. *Conservation Biology*, **12**, 1318–26.

Nature Conservancy Council (1989). *Guidelines for Selection of Biological SSSIs*. Peterborough: NCC.

Newson, M. D. (2002). Geomorphological concepts and tools for sustainable river ecosystem management. *Aquatic Conservation: Marine and Freshwater Ecosystems*, **12**, 365–79.

Noss, R. F. (2004). Conservation targets and information needs for regional conservation planning. *Natural Areas Journal*, **24**, 223–31.

Ormerod, S. J. (1999). Three challenges for the science of river conservation. *Aquatic Conservation: Marine and Freshwater Ecosystems*, **9**, 551–8.

Pringle, C. M. (2001). Hydrologic connectivity and the management of biological reserves: a global perspective. *Ecological Applications*, **11**, 981–98.

Rabe, F. W. & Savage, N. L. (1979). A methodology for the selection of aquatic natural areas. *Biological Conservation*, **15**, 291–300.

Ratcliffe, D. A. (ed.) (1977). *A Nature Conservation Review*. Cambridge: Cambridge University Press.

Raven, P. J., Fox, P., Everard, M., Holmes, N. T. H. & Dawson, F. H. (1997). River Habitat Survey: a new system for classifying rivers according to their habitat quality. In *Freshwater Quality: Defining the Indefinable?*, eds. P. J. Boon & D. L. Howell. Edinburgh: The Stationery Office, pp. 215–34.

Saunders, D. L., Meeuwig, J. J. & Vincent, A. C. J. (2002). Freshwater protected areas: strategies for conservation. *Conservation Biology*, **16**, 30–41.

Smith, R. K., Freeman, P. L., Higgins, J. V. *et al.* (2002). *Freshwater Biodiversity Conservation Assessment of the Southeastern United States. The Nature Conservancy* (www.freshwaters.org/info/large/documents.shtml).

Sowa, S. P., Annis, G. M., Diamond, D. D. *et al.* (2003). *An Overview of the Data Developed for the Missouri Aquatic Gap Project and an Example of How It is Being Used for Conservation Planning*. Gap Analysis Bulletin No, 12. USGS/BRD/Gap Analysis Program, Moscow, ID.

U.S. Fish and Wildlife Service (USFWS) (2004). Endangered and threatened wildlife and plants; designation of critical habitat for three threatened mussels and eight endangered mussels in the Mobile River Basin; final rule. Federal Register, 69 (1 July 2004), 40084–171.

Winston, M. R. & Angermeier, P. L. (1995). Assessing conservation value using centers of population density. *Conservation Biology*, **9**, 1518–27.

Wright, J. F., Moss, D., Clarke, R. T. & Furse, M. T. (1997). Biological assessment of river quality using the new version of RIVPACS. In *Freshwater Quality: Defining the Indefinable?*, eds. P. J. Boon & D. L. Howell. Edinburgh: The Stationery Office, pp. 102–8.

8 · *Methods for assessing the conservation value of lakes*

LAURIE DUKER AND MARGARET PALMER

Introduction

The identification of key areas of lake biodiversity is more critical than ever before. The projected mean future extinction rate for North American freshwater fauna is approximately five times greater than for terrestrial fauna (Ricciardi & Rasmussen, 1999). Approximately 45% of US lakes surveyed are not clean enough to support uses such as fishing or swimming (US Environmental Protection Agency, 2000). In England and Wales, of 283 lakes assessed, 138 (49%) are at risk, or are probably at risk, of failing the ecological objectives of the EC Water Framework Directive (WFD) because of contamination by nutrient phosphorus from diffuse sources (Environment Agency, 2004).

General principles

There are three basic requirements for a systematic evaluation of lakes for their nature conservation value. The first is a foundation of survey data on which to base the assessment; the second, particularly appropriate where habitats rather than species form the interest, is a site classification; the third is an analytical framework for evaluating and prioritizing sites for conservation. Sites may be assessed in local, national or international contexts. Prioritization may be based on habitat type, the importance of species or groups of species present or, more recently, on the intactness of ecological or evolutionary processes sustained by the area.

Much of the information in this chapter applies to lakes whose origin is natural. However, given the enormous number of natural lakes whose hydrology has been altered to one degree or another, naturalness itself has become a continuum. Many reservoirs now essentially replace previously existing smaller lakes, and reservoirs and small artificial waters such as gravel pits and ponds often become important substitute habitat for species

Assessing the Conservation Value of Fresh Waters, ed. Philip J. Boon and Catherine M. Pringle. Published by Cambridge University Press.

originally found in natural lakes. Although some analytical yardsticks such as naturalness of habitat (i.e. degree of departure from a pristine state) may not be appropriate for artificial water bodies, others, such as species rarity or richness, are still useful and important measures. Many artificial standing waters deserve, at a minimum, analysis as viable candidates for conservation, even if they are unlikely to be top priorities.

Biodiversity data collection and management

The analytical and prioritization process must be built on sound biological and environmental data. This raises questions about which taxonomic groups and physical features are to be targeted and which data collection methods should be used. How wide a range of features should be assessed and how much detail is required?

Traditional inventories of lake biological assets have regularly emphasized vascular plants, birds and fish, often to the detriment of smaller or less-recognized life forms such as invertebrates and plankton. However, truly comprehensive inventories of lakes, let alone lake catchments, are obviously too time-consuming and resource-intensive to be practical. How broad an inventory is optimal? Species richness in one taxonomic group does not necessarily indicate that other groups are species-rich, so decisions have to be made about which groups of organisms are most appropriate for consideration as indicators of conservation value for individual sites. In addition to species inventories, information such as animal death and deformity rates can be important indicators of lake ecosystem integrity. Predictive modelling may be used to expand the usefulness of current biodiversity data and to improve the ability to prioritize lakes for conservation. This approach is described in the section on analytical frameworks for evaluating and prioritizing sites.

Survey and data collection

In the UK, species occurrence data are collected by bodies such as statutory organizations (nature conservation agencies, Natural Environment Research Council, Environment Agency and Scottish Environment Protection Agency), NGOs (e.g. British Trust for Ornithology, Pond Conservation, Freshwater Biological Association, Botanical Society of the British Isles) and universities. There are voluntary national recording schemes, each of which deals with a particular taxonomic group of organisms.

Biodiversity data collection and management is less centralized in the USA than in Great Britain, with most work originating in state conservation agencies and non-profit organization networks at the state or catchment level. In many states, baseline information about lake species and ecological communities is gathered from sources that include scientific literature, natural history collections, expert references and past field inventories, then supplemented with additional field studies. Ideally, information is synthesized in a computer database and mapping system that can inform and clarify management decisions. Comprehensive maps of underwater lake features are also vital for informed decision-making, and The Nature Conservancy has done substantial work in this area, particularly in the Laurentian Great Lakes.

Standardizing survey methods and taxonomies, and coordinating data exchange and conservation planning across US state and national borders that artificially divide lake catchments is a significant challenge. As yet, only a handful of catchments, such as the Laurentian Great Lakes, Lake Champlain and Lake Tahoe, have appointed coordinating bodies and legislated catchment-wide cooperation and the funding to make it effective.

Approaches to species survey and data collection in the UK and the USA are summarized in Tables 8.1 and 8.2. Differences in taxonomic emphasis are evident. The taxa most used in conservation assessment are macrophytes and birds in the UK and fish and birds in the USA, probably because data are more abundant and reliable for these taxa than for others. The common emphasis on birds reflects the global popularity of ornithology and the huge resource of information contributed by keen volunteers. The UK's reliance on plant data has its roots in that country's tradition of amateur botany, boosted by an extensive lake survey programme by the statutory conservation agencies (Duigan *et al.*, 2006, 2007) and a long-running research project, the National Vegetation Classification (Rodwell, 1995). The emphasis on fish in the USA reflects the economic and recreational importance of its freshwater fisheries.

Data and information dissemination

In the UK, much of the information collected on species distribution is passed to the national Biological Records Centre, which collates it, stores data in electronic form and contributes to taxonomic atlases (Arnold, 1995; Merritt *et al.*, 1996; Preston & Croft, 1997; Davies *et al.*, 2004). Although the records are not exclusively from lakes, distribution maps give context

Table 8.1 *Survey methods and data collection*

Taxonomic group	United Kingdom	United States	References
Macrophytes	An important group for lake classification and assessment. A standard survey protocol is used, involving shore-based grapnel sampling and, where possible, boat-based transects. Relative abundance of each species is recorded using the DAFOR (Dominant, Abundant, Frequent, Occasional, Rare) scale. Statutory conservation agencies hold databases of records, currently for more than 4000 sites in Britain and Northern Ireland. Pond Conservation holds a database of macrophytes from ponds.	State biologists, researchers, lake catchment management organizations and volunteers inventory aquatic plants, map vegetation populations and quantify plant communities using line-intercept frequency, point-intercept frequency and biomass data analysis. Sweep nets and Ekman grabs are often used for sampling. Macrophytes are often inventoried to preclude or manage invasive non-native plants (see Table 8.2).	*UK* Duigan *et al.*, 2006, 2007 Wolfe-Murphy *et al.*, 1991 Williams *et al.*, 1998
Phytobenthos and phytoplankton	Core-sampling for fossil diatoms produces paleolimnological data useful for indicating man-induced changes in nutrient status or acidity. Littoral diatom assemblages are used as indicators of ecological status. Phytoplankton (e.g. through determinations of chlorophyll a and % cyanobacteria) can be used for assessing nutrient pressures.	Core sampling of sedimented diatoms is part of USEPA's proposed biological sampling methodology. Changes in species ratios can reflect changes in lake levels, water sources and eutrophic conditions. Number of taxa, diversity indices (e.g. Shannon-Weiner) and pollution tolerance indices (e.g. Lange-Bertalot) are assessed.	*UK* Bennion *et al.*, 1996 Flower *et al.*, 1997

Table 8.1 (*cont.*)

Taxonomic group	United Kingdom	United States	References
Mammals	Water shrew *Neomys fodiens*, water vole *Arvicola terrestris*, otter *Lutra lutra* and the introduced American mink *Mustela vison* are routinely recorded by the Mammal Society. Otter and water vole (protected species) have been specially targeted for survey by statutory agencies and NGOs.	Beaver *Castor canadensis* is a keystone species of particular interest. Aerial surveys effectively detect active colonies in the fall. Ground surveys may be more costly and less likely to document as high a percentage of existing colonies. More data are needed on average colony sizes in various habitats.	*UK* Strachan *et al.*, 1990 Jefferies, 2003 *USA* Murphy & Smith, 2001
Birds	Volunteer effort produces a large amount of quantitative data. Monthly UK-wide wetland bird counts are carried out through a partnership between the statutory conservation agencies and NGOs (the Royal Society for the Protection of Birds, the British Trust for Ornithology and the Wildfowl and Wetlands Trust).	Aerial surveys are widely used by the US Fish and Wildlife Service and others to assess bird populations. Many state-wide departments use volunteers to gather annual information (e.g. Minnesota Department of Natural Resources, which surveys common loons in six 100-lake regions).	*UK* www.wwt.org.uk *USA* King & Brackney, 1997 Earnst *et al.*, in review
Amphibians	Amphibia are surveyed during the breeding season by day and at night by torch-light. Frog populations are quantified by counting spawn clumps.	A new national amphibian research, monitoring and conservation initiative is led by US Geological Survey. Targeted species sampling is carried out to assess species	*UK* Arnold, 1995 *USA* Heyer *et al.*, 1994

	Warty newt *Triturus cristatus* and natterjack toad *Bufo calamita* (threatened species) are targeted, as well as species assemblages.	distributions. Methods include listening for calls, using dip-nets and looking under cover-boards.	
Fish	Fish survey often relies on fishing returns but may involve gill-netting or electrofishing. Rarer species (e.g. vendace *Coregonus albula*, Allis shad *Alosa alosa*) are specially targeted for survey for conservation purposes. The Environment Agency (England and Wales) and Fisheries Research Services and District Salmon Fisheries Boards (Scotland) are the principal repositories of data on fish stocks in Britain.	Prior to 1970 data were based mainly on commercial fish harvest records, which were influenced by demand and fishing technology. Many states now monitor fish populations and habitats more broadly and contribute to multi-state research using standardized methods. Survey methods include trap nets, gill nets, electrofishing, mark/recapture, telemetry and video or photographic documentation. Minnesota's Department of Natural Resources surveys 650 lakes each year and has a fish database covering over 4000 water bodies. Aquatic Gap pilot projects produce presence/ absence databases.	*UK* Davies *et al.*, 2004 *USA* Thoma & McNight, 1995 www.dnr.state.mn.us/lakefind/surveys.htm www.umesc.usgs.gov/data_library/fisheries/fish_page.htm American Fisheries Society, 2004
Invertebrates	In the past, lake invertebrates have not been routinely used for water quality assessment in lakes (cf. rivers). Benthic invertebrate communities are now proposed for use as indicators of acidification, and chironomid pupal exuviae have	Little emphasis has been placed on studying the distribution and abundance of invertebrates in lake habitats. Data may be used to define high quality waters, support enforcement of water quality standards and measure	*UK* Williams *et al.*, 1998 *USA* www.anr.state.vt.us/dec/waterq/lakes/htm/lp_lczebramon.htm

Table 8.1 (*cont.*)

Taxonomic group	United Kingdom	United States	References
	been shown to be reliable indicators of nutrient enrichment. The spread of invasive alien species such as signal crayfish *Pacifastacus leniusculus* and zebra mussel *Dreissena polymorpha* are monitored by the Environment Agency and other statutory agencies. The aim of invertebrate survey for conservation assessment is usually to obtain as full a species list as possible, using semi-standardized methods: sweep-netting and kick-sampling in the littoral zone; plankton netting; observation through binoculars for adult Odonata. Pond Conservation holds a database of invertebrates from many hundreds of ponds.	improvements associated with management actions. Survey methods include kick netting in littoral habitats, use of artificial substrate samplers for quantitative data and dip net/hand picked collections for qualitative samples. Distribution, abundance and new colonization of zebra mussel are monitored widely. Open water stations are examined for occurrence and density of veligers, using vertical plankton net tows. Nearshore stations more often examine settled juvenile occurrence and densities, using settling plates or bricks. Much sampling is accomplished through volunteer networks.	www.ecy.wa.gov/programs/eap/fw_lakes/lk_zebra.htm

Table 8.2 *Summary of Washington State Department of Ecology aquatic plant monitoring methods*

Objective	Suggested methods	Relative effort/cost	Data reliability
Find invasive species	Surface inventory	Low	Low
	Diver inventory	Medium	Medium
Create species list	Surface inventory	Low	Low
	Diver inventory	Medium	Medium
Detect population changes: over time assess infestations to determine appropriate control methods	Surface mapping	Medium	Medium
	Diver mapping	Medium	Medium
	Remote sensing	Medium	High
Evaluate effectiveness of control methods	Frequency of data from transects or points	Medium	High
	Biomass	High	High

Source: Parsons (2001).

to species data collected in a particular site. The availability of nation-wide species datasets also enables threat assessments to be made and Red Lists to be produced. British Red Lists have been published for numerous taxonomic groups, including vascular plants (Cheffings & Farrell, 2005). Threat assessments contribute to the formulation of national legislation and policy on nature conservation, including the UK Biodiversity Action Plan (BAP) (Department of the Environment, 1994; www.ukbap.org.uk/).

A significant initiative, being developed as part of the UK BAP, is the National Biodiversity Network (NBN) (www.nbn.org.uk). This project aims to build the UK's first comprehensive network of biodiversity information and to make data widely accessible in digitized and exchangeable form.

One of the leading national forces in assessing biodiversity in the USA is the Natural Heritage Data Center Network, a partnership between public agencies and The Nature Conservancy, a private, non-profit conservation organization. Common standards and procedures for inventory and information management are used across all independent Natural Heritage programmes. Although major inventory gaps remain for aquatic species, and more work has been done in riparian habitats than lake habitats, the Heritage Network's efforts to assess the conservation status of aquatic plants and animals in the USA and to map populations at greatest risk of extinction provide a basis for many of the lake conservation efforts initiated by state natural resources departments and non-profit organizations across the country. Often citizens groups, which tend to be less bureaucratic than state or national governments, are at the forefront of catchment-wide data collection, syntheses and planning. Developing arrangements that facilitate and institutionalize the sharing of data collection, analysis and joint planning is essential, but can take decades to evolve.

Lake classification

Lakes form a multidimensional continuum and are subject to natural stochasticity and continual disturbance. Nevertheless, broad aquatic site classifications are required under European legislation and, wherever used, they form a useful framework because they enable scientists and conservationists to

- make educated predictions about species and community distributions in areas that have not yet been thoroughly studied;
- identify priority conservation areas by allowing comparison of like with like;
- ensure that conservation priorities include the full spectrum of lake habitats;
- identify best examples (i.e. relatively pristine reference sites) of each type, to compare with other similar areas when directing and monitoring conservation efforts.

Lake classifications may be based on water chemistry, lake hydrological dynamics, aquatic habitats or lake biology. A classic approach for standing waters is to distinguish between deep, stratifying lakes, in which anoxia usually occurs below the thermocline during the summer, and shallower, non-stratifying lakes that are well mixed. Another method uses trophic categories, based on alkalinity, nutrient concentration or productivity as measured by phytoplankton biomass. There have been a number of definitions of the terms dystrophic, oligotrophic, mesotrophic and eutrophic. The Organization for Economic Co-operation and Development (1982) suggests the values in Table 8.3 in its fixed boundary classification system.

US lake catchment classifications that rely on habitat usually divide catchments into open water, shoreline, coastal wetlands, tributaries and inland habitats. Equivalent 'Phase 1' survey habitat categories in the UK would include standing water, running water, mire and swamp/marginal/inundation communities (England Field Unit, 1990). Numerous species

Table 8.3 *OECD definitions of trophic status of lakes*

	Total phosphorus	Chlorophyll *a*		Secchi disc transparency	
Trophic status	Annual mean (mg L^{-1})	Annual mean (mg L^{-1})	Maximum (mg L^{-1})	Annual mean (m)	Minimum (m)
Ultra-oligotrophic	≤ 0.004	≤ 0.001	≤ 0.0025	≥ 12.0	≥ 6.0
Oligotrophic	≤ 0.01	≤ 0.0025	≤ 0.008	≥ 6.0	≥ 3.0
Mesotrophic	0.01 – 0.035	0.0025 – 0.008	0.008 – 0.025	6.0 – 3.0	3.0 – 1.5
Eutrophic	0.035 – 0.1	0.008 – 0.025	0.025 – 0.075	3.0 – 1.5	1.5 – 0.7
Hypertrophic	≥ 0.1	≥ 0.025	≥ 0.075	≤ 1.5	≤ 0.7

of fish, invertebrate and other animals in lakes use more than one habitat for critical life stages.

Classification of lake biological communities is essential for effective conservation planning. However, the process is a long one that begins with separate evaluations for each plant and animal assemblage, many of which have widely disparate ranges and substantial overlap. Attempts to identify unique plant and animal assemblages that are found only in certain types of lakes in the USA are continuing, but meeting with limited success.

Any method of classification is fraught with problems because natural habitats, including lakes, form a continuum in space and time and partitioning is, unavoidably, a subjective business. Final decisions on where to draw the lines that separate site types are best made by experienced surveyors able to interpret data analysis in the light of their 'feel' for natural systems.

Lake classification in the UK

The lake classification systems currently used by the UK statutory nature conservation agencies are broad and largely botanical. The original British classification was based on plant records from 1124 sites (Palmer *et al.*, 1992). This has now been superseded by a classification produced by multivariate analysis of data from 3449 sites (Duigan *et al.*, 2006, 2007). The later classification recognizes 10 major lake groups (Table 8.4), representing a series lying predominantly along a gradient of alkalinity, with highly acidic sites (Group A) at one extreme and highly alkaline freshwater sites (Group I) at the other. A saline Group J obviously stands alone.

Table 8.4 *The revised British lake classification*

Lake group	Distribution and physical characteristics	'Ecotype' category	Typical vegetation
A	Mainly northern. Highly acid bog/heathland pools	Dystrophic	Dominated by *Sphagnum* species
B	Mainly northern. Acid moorland/heathland pools and small lakes	Soft water	Low diversity: *Juncus bulbosus, Potamogeton polygonifolius, Sphagnum* spp.
C1	Mainly northern and western. Acid upland lakes	Soft water	Low diversity: *J. bulbosus, Sparganium angustifolium*
C2	Predominantly north-western. Slightly acid upland lakes	Soft water	High diversity: *Littorella uniflora, Lobelia dortmanna, Myriophyllum alterniflorum*
D	Mainly northern and western. Mid-altitude circumneutral lakes	Circum-neutral	High diversity: *L. uniflora, M. alterniflorum, Callitriche hamulata, Fontinalis antipyretica, Glyceria fluitans*
E	Mainly northern. Often low altitude and coastal lakes, slightly above neutral	Circum-neutral	High diversity: *L. uniflora, M. alterniflorum, Potamogeton perfoliatus, Chara* spp.
F	Widespread. Lowland above neutral lakes	Hard water	Low diversity: water lilies and other floating-leaved vegetation
G	Mainly central and eastern. Above neutral lowland lakes	Hard water	*Lemna minor, Elodea canadensis, Potamogeton natans, Persicaria amphibia*
H	Northern. Small, circumneutral, lowland	Hard water	Low diversity: *G. fluitans, Callitriche stagnalis*
I	Widespread. Base-rich lowland lakes	Hard water with *Chara*	*Chara* spp., *Potamogeton filiformis, P. pectinatus, Myriophyllum spicatum*
J	Northern. Coastal brackish lakes	Brackish	Low diversity: *P. pectinatus, Ruppia maritima, Enteromorpha*, fucoid algae

Source: Adapted from Duigan *et al.* (2006, 2007).

The Northern Ireland lake classification (Wolfe-Murphy *et al.*, 1992, Heegaard *et al.*, 2001) recognizes 16 lake types in four groups: base-poor upland lakes, man-made water bodies, two types of white water lily *Nymphaea alba – Fontinalis antipyretica* assemblage and seven types of yellow water lily *Nuphar lutea – Elodea canadensis* assemblage.

One problem with these botanical classifications is that data are from both near-natural and impaired sites, and the analyses do not satisfactorily distinguish them. Site-specific methods of determining whether water chemistry has been substantially modified include paleolimnological studies (Bennion *et al.*, 1996) or 'hind-casting', using statistics on past catchment land use to establish a baseline state for nitrogen and phosphorus concentrations, for comparison with current concentrations (Moss *et al.*, 1996).

Lake classification in the USA

Unlike the situation in Great Britain, where one classification system has been devised for the entire country, lake classification work in the USA has been mainly at the state level. State boundaries arbitrarily divide aquatic biological assemblages, complicating the already difficult task of lake conservation planning. Institutional arrangements between states are necessary, to build accurate catchment-wide approaches to each step of the process in cross-boundary lake catchment conservation. To date, most aquatic classification work in the USA has centred on riparian habitats. Some attempts have been made to develop aquatic classifications of biological communities in tributaries based on the terrestrial characteristics of the drainage (Rahel & Hubert, 1991; Lyons, 1996).

Portions of certain national US laws guide or inform aspects of lake conservation, resulting in some similarities in lake classification and conservation planning across the country. The Clean Water Act, which was passed in its modern form in 1972, is the cornerstone of surface water quality protection in the USA (see Chapter 2). The statute sets a goal of restoring and maintaining the 'chemical, physical, and biological integrity of the nation's waters'. Recently, renewed emphasis has been placed on the 'biological integrity' portion of the goal, resulting in increased focus on aquatic biological data collection, lake classification and ultimately analysis and prioritization of lake conservation targets.

Many state coalitions, usually comprising state natural resource departments and local, state and national non-profit organizations, are attempting to identify biological assemblages that include macrophyte, macroinvertebrate, fish and/or amphibian species, in order to identify and understand intact lake ecosystems in a broader sense, hopefully leading to more effective and comprehensive lake conservation.

The Nature Conservancy and the Vermont Biodiversity Project have jointly developed six separate aquatic classification schemes (Langdon *et al.*, 1998), each associated with one or more biological groups:

- macrophytes of standing waters;
- macroinvertebrates of standing waters;
- macroinvertebrates of running waters;
- fish of standing waters;
- fish of running waters;
- reptiles and amphibians.

Classification is based on currently available biological data housed in extensive biological databases on fish, macroinvertebrates, aquatic plant, and amphibian and reptile communities. Biological assemblage types reflect similarities in species occurrence data between near-natural reference sites. Where biological data are unavailable or incomplete, aquatic biologists create conceptual categories using physico-chemical habitat types as surrogates.

More work is needed on classifying open-water plant and animal communities. As a result of costly data collection methods, data are often lacking. Despite the importance of underwater plants in the Laurentian Great Lakes, for example, relatively little is known about them. In some areas, underwater plants may be the dominant primary producer in the food chain supporting animal populations.

Analytical frameworks for evaluating and prioritizing lake conservation sites

In the same way that biodiversity exists at many levels (genes, species, communities, ecosystems and landscapes), there are different biogeographical scales, such as individual sites, lake catchments, river basins and ecoregions, that must be taken into consideration when prioritizing lake conservation targets. Broad-scale approaches ensure that important processes and distribution patterns that may have cumulative impacts are identified. Broad-scale approaches also provide important information for those who develop policy and law. At finer scales, analysis is critical for site-specific projects (Regional Ecosystem Office, 1995).

Evaluation and prioritization techniques in the UK

Nationally important sites

Nationally important sites (Sites of Special Scientific Interest (SSSIs) in Britain and Areas of Special Scientific Interest (ASSIs) in Northern Ireland) are designated and enjoy legal protection (see Chapter 2). *Guidelines for Selection of Biological SSSIs* (Nature Conservancy Council,

1989), although now in need of updating, remains a seminal work on site evaluation and selection in Great Britain. Its rationale is built on the philosophy of *A Nature Conservation Review* (Ratcliffe, 1977), and it contains detailed guidelines both for habitats and for species. Administrative areas of the statutory nature conservation agencies in England, Scotland and Wales are used as areas of search in these guidelines, but 'natural areas' (Usher & Balharry, 1996) are now becoming accepted as more logical areas of search. Sites selected within each of these areas contribute to a national network of protected SSSIs, covering all near-natural habitats and taxonomic groups. A minimum aim is to represent all the important habitats and species that occur within each area by at least one, and preferably the best, example or population. The *Guidelines* make recommendations for minimum standards to be expected in SSSIs, but the final decision on site selection depends on expert judgement. The chapter in the *Guidelines* on standing waters uses the earlier British classification of lakes (Palmer *et al.*, 1992). It recommends consideration of the following criteria where sites are being selected on the grounds of habitat:

Naturalness This criterion encompasses anthropogenic nutrient enrichment and acidification, as well as intactness of structure. Artificial structures and introduced plants are negative features; naturalness of the catchment is a positive feature. Unspoilt palaeolimnological and geological features add value.

Typicalness The full range of site types present in each area of search should be represented in the SSSI series. No guidance is given on distinguishing naturally nutrient-rich systems from artificially enriched ones, but in the light of more recent work (Bennion *et al.*, 1996; Moss *et al.*, 1996; Flower *et al.*, 1997) this would be an important addition to any revision of the *Guidelines*. The site classification also needs bringing up to date.

Rarity Unusual site types, for instance machair lochs, arctic-alpine lakes, brackish lochs or aquifer-fed, naturally fluctuating waters, should receive special consideration. Sites of a type rare in a particular part of Britain, such as acid waters in southern and eastern England, should be well represented. The presence of rare species can contribute to this criterion, especially if the species are recognized as rare in international statutes and conventions.

Diversity Generally, only sites with at least the average number of species for the relevant site type should be considered for selection. The presence of eight or more species of *Potamogeton*, at least one of which is nationally or locally rare, would qualify a site for consideration. Diversity of physical features should be taken into account.

Position in an ecological unit Where an ecological series exists (e.g. a chain of lakes spanning a natural transition from nutrient-poor to nutrient-rich conditions) it is important to include the full range of water bodies within SSSI boundaries. Intactness of connectivity is important where this criterion is applied.

Size Large sites may be preferable to small ones where diversity or the population size of an important species depends on extent. However, large sites usually have extensive catchments, which may pose conservation problems. Where a site type is well represented in an area of search, a range of sites of various sizes may be selected.

In Northern Ireland, less detailed guidance (Environment and Heritage Service, 1999), based on the same principles as the SSSI Guidelines, has been issued for selecting ASSIs within the framework of the sixteen site types recognized in the lake classification.

In *Guidelines for Selection of Biological SSSIs* the approach to site selection for species varies from one taxonomic group to another, but rarity, species-richness, naturalness and population size are emphasized. The *Guidelines* recommend that well-established, rather than ephemeral, populations should be considered, that SSSIs should be large enough to support viable populations, that weight should be given to representation of species at their geographical limits and that protected sites should form an interdependent national network that facilitates interchange between populations. The following examples illustrate the approach.

Birds Site evaluation criteria for birds are based largely on population size, the threshold being 1% or more of the British breeding or non-breeding population. Also eligible for selection are localities supporting an especially wide range of bird species characteristic of the habitat, or localities with recent records for at least 70 breeding species, at least 90 wintering species or at least 150 species on passage. SSSIs may also be designated for rare bird species and unusual features, such as inland cormorant colonies.

Amphibians For amphibians, the presence of strong populations of at least four breeding species qualifies a site for consideration. All well-established breeding sites of the rare natterjack toad *Bufo calamita* are eligible for selection, also all sites where a night count of warty newt *Triturus cristatus* in the breeding season exceeds 100 individuals.

Odonata The overall aim of the *Guidelines* for dragonflies and damselflies is to represent within the SSSI series every species established in the area of search, paying particular attention to breeding sites of internationally and nationally threatened species. All sites that support outstanding assemblages are candidates for selection. The numbers of species cited as constituting outstanding assemblages vary from 17 in the south of England to 7 in the Orkney Islands, but these figures need revising in the light of recent analysis of records.

The ideal is an integrated approach, using general habitat characteristics and a variety of taxonomic groups, as has been done for the suite of lakes in Anglesey, Wales (Duigan *et al.*, 1996), but a sufficiently wide range of data is not always available.

Wetscore

A scoring system for evaluating and ranking sites for water beetles has been developed (Foster *et al.*, 1990) and dubbed 'Wetscore'. This method assigns a quality score to each species, related to its national and regional rarity. Scores are allotted on a six-point geometric scale, with very common species scoring 1 and the most threatened Red List species scoring 32. A Species Quality Index (SQI) for the site is calculated by dividing the total score for the site by the number of species found. SQIs for sites of a similar type can then be compared to provide a numerical measure of the value of sites for their water beetle assemblages.

Lake Habitat Survey (LHS) and Lake Assessment for Conservation (LACON)

Two techniques for lake evaluation currently under development in the UK are Lake Habitat Survey (LHS) (Rowan *et al.*, 2006a,b) and Lake Assessment for Conservation (LACON) (Palmer, in prep.). The two systems are closely related and are evolving in tandem.

A standard, European-wide habitat survey methodology is required for assessing both the condition of lakes designated as Special Areas of Conservation and the intactness of lake hydromorphology under the WFD (Chapter 2). LHS is being developed to fulfil these needs. The protocol covers

- the collection of background information (e.g. lake area, perimeter, depth and altitude; catchment characteristics);
- field survey of physical attributes of the riparian, shore and littoral zones, including vegetation structure;
- an overview of the hydrology of the lake;
- an assessment of human pressures on the site.

Indices of habitat quality and the extent of lake modification have also been developed.

LACON is a British system based on SERCON principles for rivers (see Chapter 7). LACON is not a replacement for *Guidelines for Selection of Biological SSSIs*, rather it is a method for identifying possible candidate sites for further assessment. LACON elaborates the method recommended in the *Guidelines* by using a range of attributes under each criterion and applying a weighted scoring system to achieve rigour and repeatability in the assessment. Five conservation criteria are scored in LACON: physical diversity, naturalness, representativeness, rarity and species richness. Impacts are also scored. So far, LACON only covers plants in any detail. Scores (or indices) for each criterion are kept separate, but the final assessment takes into account the whole range of indices. A site might score highly for diversity but have a low naturalness index, and the final evaluation would depend on expert judgement and the context of the site. LACON incorporates reference data on which the scoring is based, for instance checklists of native and introduced aquatic plant species. There are plans to automate the scoring system.

Physical diversity In LACON three attributes of physical diversity are currently considered: the number of substrate types, the variety of macrophyte growth forms and 'other' habitat features such as types of shoreline, inflows and outflows, islands and bays. The categories are closely related to those in LHS.

Naturalness Three attributes of naturalness are scored: physical structure, aquatic vegetation and adjacent habitat. Naturalness of physical structure is assessed by considering the origin of the lake basin, modifications to the hydrological regime and the extent of artificial structures on its banks. Departure from naturalness for aquatic vegetation is assessed using the presence and abundance of introduced species, both plants alien to Britain and those introduced from another part of the country. Naturalness of adjacent habitat is evaluated by estimating the percentage

Figure 9.1. Distribution of lakes in Sweden.

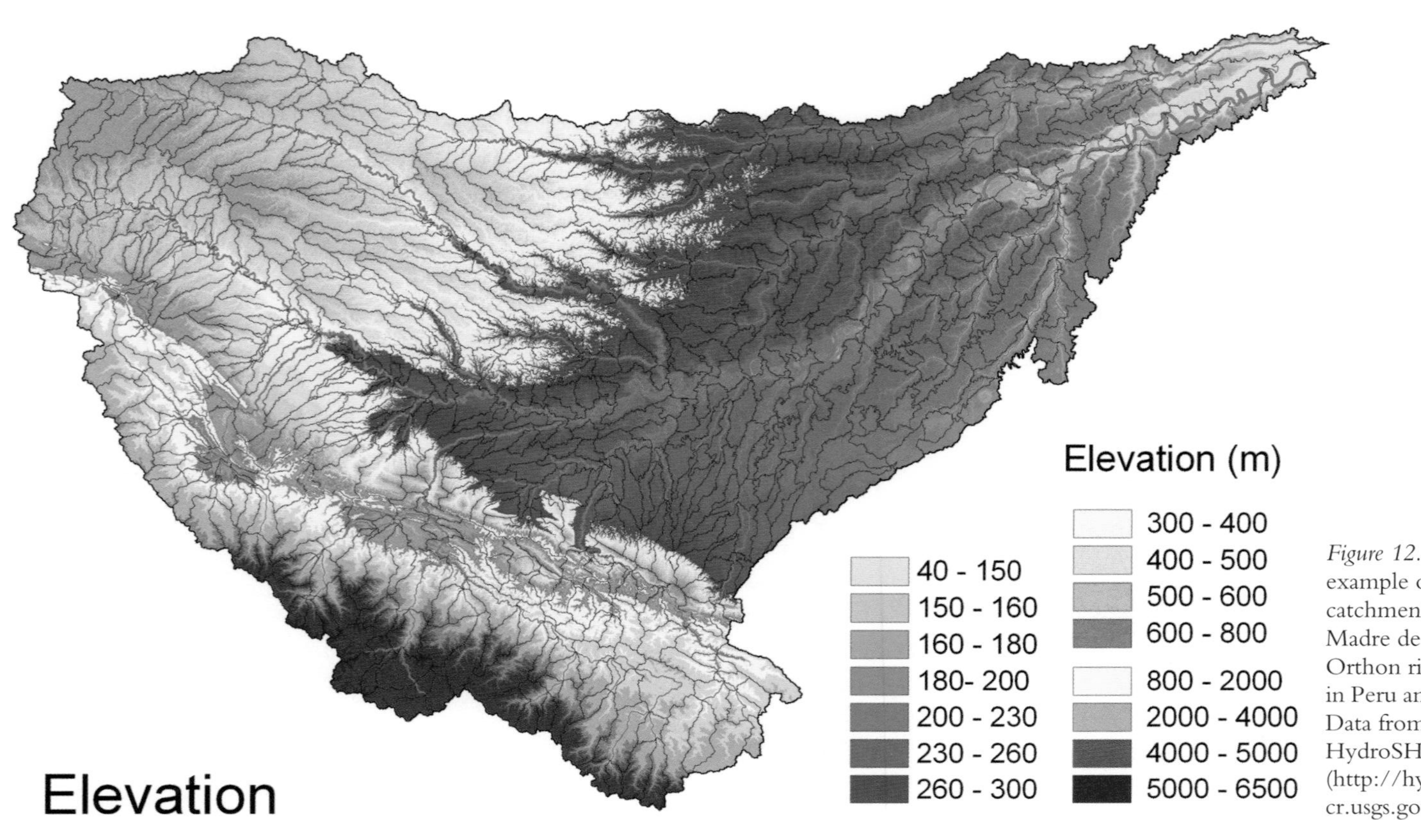

Figure 12.1. An example of catchments for the Madre de Dios and Orthon river basins in Peru and Bolivia. Data from HydroSHEDS (http://hydrosheds.cr.usgs.gov/).

of the lake perimeter abutted by natural habitats such as native woodland, marsh, heathland or unimproved grassland.

Representativeness Representativeness is assessed by comparing the macrophyte species list for a site with a list of typical species for the relevant type of water body, derived from the revised botanical classification (Duigan *et al.*, 2006, 2007). A second method uses the mean Plant Lake Ecotype Index (PLEX) for the site. PLEX values for individual macrophyte species have been calculated (Duigan *et al.*, 2006), indicating strength of association with five broad freshwater 'ecotype' categories (see Table 8.4). A mean PLEX score for a site is calculated, using the PLEX values for the species recorded there. A measure of representativeness is obtained by comparing the mean PLEX score for the site with the range for all the sites in the relevant lake group in the national database.

Rarity The occurrence of rare and protected aquatic plant species is the basis for the LACON assessment of rarity. The categories form a hierarchy, ranging from internationally and nationally protected species, to Red List species, to nationally localized species, to locally uncommon ones. The higher the category the more weighting is given to the score.

Species richness Species richness is assessed by counting the number of native aquatic vascular plant, charophyte and bryophyte species recorded in the lake and the number of swamp and tall-herb fen communities (Rodwell, 1995) present.

Impacts This section of LACON deals with the impact of human activity on the macrophyte flora of lakes and uses similar categories to those in LHS. Scores under *Impacts* provide context for the conservation criterion scores and can help in the development of appropriate management strategies. The strength of the impact is assessed for acidification; eutrophication; water abstraction and water level management; recreational and educational use; introduced species; modification of the lake basin; and other pressures from surrounding land use.

Additional features of importance These features in LACON are not scored, but they can be influential in the overall assessment of the value of a site. Positive features include extensive hydroseral development, standing waters of a kind rare internationally (e.g. brackish lagoons, turloughs) and water bodies that occupy a key position in an ecological series (e.g. the transition from fresh to brackish water). Negative features include factors that are not assessed in the 'Impacts' section, such as pollution affecting biota other than plants.

The Predictive System for Multimetrics (PSYM)

PSYM has been developed by the Environment Agency and Pond Conservation for assessing the biological quality of still waters (Williams *et al.*, 1998). It is based on survey data from 300 ponds and small lakes in England and Wales, covering a wide range of altitudes, land types and degrees of site impairment. PSYM combines a predictive approach with a multimetric-based method for ecological quality assessment to indicate the extent of departure from reference (unimpaired) condition. The PSYM computer program calculates metrics (see Table 8.5) from the biological survey data and uses simple environmental data (e.g. altitude, geology, area, pH, shade) to predict scores for the site. An Ecological Quality Index (EQI) for each metric is produced by dividing the observed scores by the expected scores. EQI is transformed to a four-point (0–3) scale or Index of Biotic Integrity (IBI). Finally, IBIs for all the metrics are summed to give an overall IBI, expressed as a percentage of the maximum possible score, which represents unimpaired condition.

A comparison of site quality assessments using LACON and PSYM

Table 8.5 shows the results of applying LACON and PSYM to data from a pond in Castor Hanglands National Nature Reserve, Cambridgeshire (Figure 8.1), which was surveyed for plants and invertebrates in 1971/72 (Palmer, 1973). The LACON assessment produced average scores for naturalness and representativeness, a high (although not outstanding) score for plant rarity and a moderately high score for macrophyte species richness. PSYM scores indicate that the pond was exceptionally rich in plant species, both common and uncommon, above average in its scores for invertebrates, and unimpaired. The survey in 1971/2 involved several visits to the pond, so the PSYM scores are somewhat higher than might have been the case for the standard single visit and 3-minute invertebrate pond net sample. However, a resurvey in 1992 using the standard PSYM methodology indicated that the site was still unimpaired and attained the maximum possible IBI (Pond Conservation, pers. comm.).

This comparison of evaluation methods illustrates both the need to appreciate details of the attributes used in the assessment and the importance of context. Unlike LACON, PSYM establishes reference

Table 8.5 *Site quality assessments using LACON and PSYM: 1971/72 data from Castor Hanglands pond, Cambridgeshire, UK*

LACON

Attributes	Indices: National mean*	Indices: National range*	Indices: Castor Hanglands	Comment
Naturalness	86	20–100	85	Spring-fed artificial pond with mainly natural margins. Surrounding habitat natural. No introduced plant species. Aquatic vegetation highly natural despite artificial basin.
Representativeness (vegetation)	69	0–100	70	3 species expected to occur in Lake Group G absent from pond. PLEX within expected range. Relatively representative of type.
Rarity (plants)	17	0–83	36	Two Red List vascular plant species. Considerable rarity value.
Species Richness (aquatic plants)	41	0–93	53	20 species of aquatic plant. 4 National Vegetation Classification swamp/tall-herb fen communities. Moderately species-rich.
Additional Features of Importance	N/A	N/A	N/A	Pond in a National Nature Reserve. Rich assemblages of dragonflies and amphibians. Overall value high because pond a significant part of a complex of habitats in the Reserve.

PSYM

Metric	Observed number/ score	Predicted number/ score	Ecological Quality Index	Index of Biotic Integrity	Comment
Plants					
Number of submerged + marginal species	56	18.4	3.0	3	Exceptionally high number of wetland species
Number of uncommon plant species	16	3.1	5.2	3	Very large number of uncommon plants

Table 8.5 (*cont.*)

PSYM					
Metric	Observed number/ score	Predicted number/ score	Ecological Quality Index	Index of Biotic Integrity	Comment
Trophic Ranking Score**	8.6	8.6	1.0	3	As predicted (for a calcareous site)
Invertebrates					
Average Score per Taxon***	5.4	5.1	1.1	3	Slightly more uncommon species than predicted
Number of Odonata + Megaloptera families	5	3.3	1.5	3	12 breeding Odonata species
Number of Coleoptera families	4	3.7	1.1	3	Slightly richer than predicted
Overall assessment					
Sum of IBI scores				18	Maximum score
% of maximum IBI score				100	An unimpaired site: unusually high score

*Based on testing the LACON system on 70 standing waters in England, Scotland and Wales.
**Described in Palmer *et al.* (1992).
***Based on sensitivity to pollution as measured by the Biological Monitoring Working Party (BMWP) Score (Armitage *et al.*, 1983).

conditions based on unimpaired sites. In LACON the plant checklist covers only fully aquatic vascular plants, bryophytes and charophytes, whereas in PSYM the checklist is heavily biased towards vascular plants, but includes a large number of wetland species as well as aquatics. The context for LACON is the whole gamut of British standing waters, ranging from large upland Scottish lochs to small artificial water bodies in lowland England, whereas the sites in the PSYM database are standing waters no bigger than 5 ha in area in England and Wales. Judged by the standards of LACON, Castor Hanglands pond would not be of national importance in itself, but would form a valuable contribution to the larger complex of the National Nature Reserve. However, PSYM

Figure 8.1. Castor Hanglands pond, Cambridgeshire, UK, in the early 1970s. © Natural England.

would place the pond amongst the richest and least impaired in England and Wales.

Evaluation and prioritization techniques in the USA

Numerous approaches are used in the USA to assess the conservation value of lakes. Plant and animal biota, their unique life histories, and the environmental conditions, biotic interactions and ecological and evolutionary processes that sustain them, all require careful assessment, as do threats both from within the catchment and beyond.

In contrast to the UK, where the approach to assessing lake conservation priorities is quite centralized, priorities for lake and aquatic species conservation in the USA are informed by federal law, but most often carried out at the state and local level. Although federal land management agencies are beginning to mandate catchment-based analyses, there is essentially no federal effort to prioritize national lake resources for conservation, a possible exception being a large programme aimed at coordinating federal work to protect and restore the Laurentian Lakes, a de facto top priority.

Conservationists use what information is available from data collection and lake classification efforts to prioritize lake conservation areas, using a range of approaches that tend to combine such factors as

- naturalness of habitat, including flow regimes;
- ecosystem processes such as migration and floodplain dynamics;
- species and habitat rarity;
- species and habitat representativeness;
- species richness and diversity;
- a lake's placement in the mountain to sea chain;
- gaps in protected areas;
- threat analysis.

Freshwater ecoregions (World Wildlife Fund – USA)

Ecoregions are 'relatively large areas of land or water that contain a geographically distinct assemblage of natural communities, and share a large majority of their species, dynamics and environmental conditions' (Dinerstein *et al.*, 1995). The World Wildlife Fund–United States (WWF–US), with support from the US Environmental Protection Agency, has conducted a landmark study identifying freshwater ecoregions of North America and analysing which features of which ecoregions should be protected or restored first (Abell *et al.*, 2000). For their analysis, WWF uses a habitat representation approach, which emphasizes the importance of conserving a full representation of diverse habitats and ecosystems as markers for freshwater biodiversity that has not been fully studied (Noss & Cooperider, 1994). WWF uses the distributions of fish, crayfish and mussels (all obligate freshwater inhabitants over their whole life cycle) to delineate initial freshwater ecoregions, in addition relying on a US Department of Agriculture (USDA) Forest Service mapping project which was based on native fish distributions.

The continental USA, Canada and Mexico are divided into 76 freshwater ecoregions, approximately 41 of which encompass some portion of the USA. Prioritizing these ecoregions is based on achieving 'the fundamental goals of biological conservation: (1) representation, (2) sustaining viable populations, (3) maintaining ecological processes, and (4) responsiveness to short- and long-term change' (Noss, 1991; Abell *et al.*, 2000).

The WWF freshwater ecoregions approach synthesizes information about the ecoregion's biodiversity (termed 'biological distinctiveness') and its naturalness (termed 'conservation status'). The main aspects of each component are described in Table 8.6.

Table 8.6 *Elements of biological distinctiveness and conservation status utilized in WWF-US freshwater ecoregion analysis*

Biological distinctiveness	Conservation status
Species richness	Degree of land cover alteration in catchment
Distinctiveness of species, communities and habitat types	Water quality degradation
Diversity at higher taxonomic scales (e.g. genus, family)	Alteration of hydrographic integrity
Rarity of ecological or evolutionary processes	Degree of habitat fragmentation
	Additional losses of original intact habitat
	Effects of introduced species
	Direct species exploitation

Source: Data drawn from Abell *et al.* (2000).

Based on their analysis and synthesis of biological distinctiveness and conservation status data, WWF–US identifies five ecoregions as top priority, two of which are in the USA. The Teays-Old Ohio or the Tennessee-Cumberland ecoregions are temperate headwater and lake habitats. Numerous lakes are identified as highly biodiverse and high conservation priorities in the Priority II category, including Lake Superior, Great Salt Lake, Bear Lake, Utah Lake, Sevier Lake and Lake Waccamaw. All the ecoregions in the Priority II category are globally or continentally outstanding in terms of biological distinctiveness.

Mapping species occurrences

Many projects use presence/absence data to map and assess lake biodiversity and identify priorities. State-agency-based Natural Heritage Network and The Nature Conservancy (TNC) assess the conservation status of plants and animals and map population occurrences of species at greatest risk of extinction. Mapped species are categorized as imperilled, vulnerable or listed under the federal Endangered Species Act as threatened or endangered (US Environmental Protection Agency, 1999; EPA Watershed Information Network, www.epa.gov/win/). Many inventory gaps for aquatic species remain. Researchers at the Ohio Aquatic Gap pilot project used their presence/absence database of fish, crayfish and mollusc species in the Lake Erie region to identify 14-Digit Hydrologic Units as high-priority conservation areas in which all three taxa were present at the 75th percentile (American Fisheries Society, 2004).

Functionality versus biological integrity

Although The Nature Conservancy's Freshwater Initiative (FWI) staff and partners understand the importance of ecological integrity, they advocate a more pragmatic standard for the conservation of lake biodiversity they call 'ecosystem functionality' (The Nature Conservancy, 1999).

There have been changes over time in the conservation approaches taken in the United States towards aquatic biodiversity conservation. The shift in approach from protecting individual rare or endangered species towards protecting overall biodiversity in key habitats in the USA is both important and fraught with some potential dangers. Trying to protect every species in every habitat and restore to full 'naturalness' is not practical, and may leave smaller, critical steps undone. The goal of biological integrity, or 'the ability to support and maintain a balanced, integrated, adaptive community of organisms having a species composition, diversity, and functional organization comparable to that of natural habitat of the region' (Angermeier & Karr, 1994), may be both impractical and a poor use of resources. Functionality, on the other hand, is a more pragmatic standard that describes the 'ability of a conservation area to sustain conservation targets and supporting ecological processes within their natural ranges of variability' (Poiani *et al.*, 2000). Functionality is translated into four attributes to evaluate the status of a freshwater conservation area at any scale:

- composition and structure of the biodiversity targets
- environmental regimes and natural disturbances
- minimum dynamic area
- connectivity.

The functionality approach provides a useful framework for assessing how intact a given lake habitat is, and may help define specific, broad catchment objectives that will support healthy biodiversity.

Federal initiatives

Protection for specific aquatic species is guided in the USA by the controversial Endangered Species Act, first signed into law in 1973, and later strengthened to include all species and to make it a federal offence to 'take' any listed species (see Chapter 2). Work done by non-profit networks and state natural resource departments on establishing priority species and habitat priorities tends to run far ahead of the US Fish and Wildlife Service's ability to use the data to identify endangered species and critical habitat for their recovery. The listing process has been contentious and slow. Larger, more visible species of lake life tend to garner more attention.

Global, national and regional approaches to lake biodiversity assessment and conservation

There are various treaties (conventions), designations and programmes relevant to assessing lakes for their conservation value. Some are international, others operate at a national or regional level. Global treaties and programmes play important roles in assessing lake biodiversity. They are also vehicles for educating government officials and the general public, and can help to draw attention to endangered lake species and habitats. Treaty secretariats and forums provide opportunities for discussing effective methods and approaches and coordinating activities across national boundaries.

The following section provides a brief description of some of these, with further information given in Chapter 2.

Ramsar Convention

The Ramsar Convention is the only global intergovernmental treaty to address explicitly the conservation of lake biodiversity. Sites designated under this Convention are selected because of their international significance in terms of ecology, botany, zoology, limnology or hydrology. They form a global network spanning a wide range of wetland habitats. At present there are 605 sites that include more than 8 ha of permanent freshwater lake habitat, and 172 of them more than 8 ha of seasonal or intermittent freshwater lake habitat. The UK has 168 designated Ramsar sites, covering about 920 000 ha, and over 80 proposed sites. Sixty designated sites contain more than 8 ha of freshwater lake habitat. In the USA there are only 24 Ramsar sites, but they extend to 1.3 million hectares, and half of them contain more than 8 ha of permanent freshwater lake habitat.

European Directives

The Habitats Directive and the Birds Directive apply within the European Union and, amongst other things, require the selection and designation of Special Protection Areas for birds and Special Areas of Conservation for threatened habitats listed on Annex I of the Habitats Directive, and for a

Table 8.7 *Habitats Directive Annex I habitats with their nearest equivalents in the British lake classification*

Annex I habitat type	GB lake groups
Oligotrophic waters containing very few minerals of sandy plains *(Littorelletalia uniflorae)*	B, C
Oligotrophic to mesotrophic standing waters with vegetation of the *L. uniflorae* and/or of the *Isoëto-Nanojuncetea*	C, D, E
Hard oligo-mesotrophic waters with benthic algae of *Chara* species	E, F, I
Natural eutrophic lakes with *Magnopotamion* or *Hydrocharition*-type vegetation	E, G, I
Natural dystrophic lakes and ponds	A
Coastal lagoons	J
Mediterranean temporary ponds	Not represented
Turloughs	Not represented

Source: Adapted from Duigan *et al.* (2006, 2007).

range of animals (other than birds) and plants that are listed on Annex II of the Habitats Directive. Areas designated under these two directives are called Natura 2000 sites. There are now nearly 25 000 of these sites, covering 17% of the land surface of the European Union.

The Annex I standing water habitats that occur in the UK, with their nearest equivalents in the British lake classification scheme, are shown in Table 8.7. Site assessment criteria for Annex I habitats and Annex II species are given in Chapter 2 (Table 2.1).

Important Plant Areas

An Important Plant Area (IPA) (Palmer & Smart, 2001) is a non-statutory site designation indicating an area of exceptionally high botanical value, judged in an international context. The three criteria for selecting an IPA are:

A. The site holds significant populations of one or more species that are of global or European conservation concern.
B. The site has an exceptionally rich flora in a European context in relation to its biogeographic zone.
C. The site is an outstanding example of a habitat type of global or European plant conservation and botanical importance.

Under the auspices of the NGO Plantlife, a list of IPAs in the UK has been drawn up for stoneworts (charophytes) (Stewart, 2004). These complex aquatic algae are acknowledged as indicators of good water quality. Of the 118 best UK sites identified for stoneworts, 38 are thought to be of European importance. These include natural lakes such as Upper Lough Erne in Northern Ireland and machair lochs in the Outer Hebrides; coastal lagoons such as The Fleet in Dorset; dune slack ponds (e.g. at Sefton dunes, Merseyside); temporary ponds in the New Forest, Hampshire; and artificial standing waters, including the Norfolk Broads and flooded brick pits near Peterborough, Cambridgeshire.

The Water Framework Directive

The WFD (Chapter 2), adopted by the European Union in 2000, aims to improve the aquatic environment in Europe. Member States are required to take steps to protect water bodies at high or good ecological status and to return to good those that are of moderate, poor or bad status. Initial stages in implementing the WFD have included the development of typologies, and protocols for classifying and monitoring ecological status.

Lake typology is based primarily on geology and depth in Britain and on geology, depth and altitude in Northern Ireland. Moss *et al.* (2003) have produced and tested a pan-European typology and classification system for shallow lakes, as a contribution to the debate on the implementation of the Directive.

Western Hemisphere Shorebird Reserve Network

In 1985 the Western Hemisphere Shorebird Reserve Network (WHSRN) was established to protect many of the key habitats across the Americas used by migrating shorebirds that undertake long journeys from northerly breeding areas to wintering grounds further south. In 2003 the network of more than 165 organizations included 58 officially recognized sites in seven countries, 8 of which were lakes in the USA, as detailed in Table 8.8. US lakes in the WSHRN network were identified as critical habitats for migratory aquatic birds only through years of coordinated international research, underscoring the importance of continued collaboration across national boundaries.

Table 8.8 *Western Hemisphere Shorebird Reserve Network member lakes in the United States*

Type of Site	Definition	Lake included in site
Hemispheric	Having at least 500 000 shorebirds	Great Salt Lake
International	Having at least 10 000 shorebirds	Mono Lake
		South Texas
		Salt Lakes
		Long Lake
		Swan Lake
Regional	Having at least 20 000 shorebirds	Benton Lake
		Salton Sea
		Lake Erie marshes

Source: Data drawn from WHSRN website (www.manomet.org/WHSRN/).

International Red Lists

Other worldwide biodiversity assessments, such as the Red List created by the IUCN Species Survival Commission, have identified aquatic species that are globally or regionally threatened (www.redlist.org/). Native UK species associated with lakes and included in the current global Red List are white-clawed crayfish *Austropotamobius pallipes* (Vulnerable) and freshwater pearl mussel *Margaritifera margaritifera* (Endangered), as well as five wetland bird species that occur as vagrants. In the USA the Red List currently identifies 12 lake species that are globally Endangered or Vulnerable. Species identified as Endangered include the whooping crane *Grus americana*, the marbled murrelet *Brachyramphus mamoratus*, the black-spotted newt *Notophthalmus meridionalis* and the smalltooth sawfish *Pristis pectinata* (www.redlist.org/search/search-basic.php).

LakeNet

In 2001, LakeNet, a global network of people and organizations, identified 250 lakes in 73 countries as initial priorities for biodiversity conservation, based on available data on fish, mollusc, crab, shrimp and bird biodiversity supported by each lake, and the rarity of certain representative types of lakes (Duker, 2001). The list represented a synthesis of Ramsar, WWF and World Conservation Monitoring Centre (WCMC) major studies on lakes, with expert input from Network advisers. In 2003, the initial list of 250 priority lakes was revised, based on feedback and newly available information from the Ramsar Bureau, LakeNet's global network of individuals and organizations in over 90 countries, and new WWF regional aquatic biodiversity studies. The 2003 revision seeks to begin to rectify the bias in the first study towards European lakes (where more key lakes have been designated as Ramsar sites) and the relative dearth of lakes in developing countries (owing to insufficient biodiversity data and lower numbers of sites nominated as Ramsar wetlands). The revised LakeNet study highlights two lakes in Scotland with outstanding biodiversity (Cairngorm Lochs and Loch of Strathbeg) and 15 in the USA (www.worldlakes.org/maps.asp?geotypeid=3, 2004).

Comparison of lake biodiversity assessment methods in the UK and the USA

Important and impressive work is being done in both the UK and the USA to collect lake biodiversity information and to apply increasingly thorough analytical methods and overarching, realistic frameworks to the data, to facilitate prioritization of lake conservation targets.

Lake biodiversity assessment in the UK appears to reap great benefits from centralization, in terms of having national databases and systems of analysis applied broadly across the entire area. While plants and birds may be good indicators of some types of biodiversity, the need to lay more emphasis on other taxa in site analysis is clear. Implementation of the WFD should encourage this approach.

In both the USA and the UK, effective volunteer monitoring programmes and non-profit networks are playing important roles in lake biodiversity assessment and provide models that may be transferable to other countries. The focus at the national level on water quality standards over more comprehensive biodiversity protection and functional integrity approaches is shifting, with countries and states looking at innovative approaches to assessing and conserving important lake habitats.

Conservationists in both countries need to continue to broaden their work to encompass a wider view of diversity that incorporates more of the species within lake ecosystems, including those of associated wetland habitats. In the USA, data collection, data analysis and lake management must be tackled at the catchment, instead of the political jurisdictional level. The WFD should promote this in the UK, as planning and management of the aquatic environment will be carried out at the river basin level. Both the USA and the UK also need to continue working with those beyond national borders to ensure the continuation of research and conservation efforts. This is especially important in catchments that cross borders and for migratory species, many of which use lakes and the waters that connect them as lifelines in their remarkable journeys.

References

Abell, R., Olson, D. M., Dinerstein, E. *et al.* (2000). *Freshwater Ecoregions of North America: A Conservation Assessment.* Washington, DC: Island Press.

American Fisheries Society. (2004). *Ohio Aquatic GAP: Assessing Fish, Crayfish and Bivalves in Relation to Conservation Lands.* Poster presented at the American Fisheries Society (http://oh.water.usgs.gov/ohgap/afs_final_poster.pdf).

Angermeier, P. L. & Karr, J. R. (1994). Biological integrity versus biological diversity as policy directives. *BioScience*, **44**, 690–7.

Armitage, P. D., Moss, D., Wright, J. F. & Furse, M. T. (1983). The performance of a new biological water quality score system based on macro-invertebrates over a wide range of unpolluted running-water sites. *Water Research*, **17**, 333–47.

Arnold, H. R. (1995). *Atlas of Amphibians and Reptiles in Britain.* ITE Research Publication No. 10. London: Her Majesty's Stationery Office.

Bennion, H., Duigan, C. A., Haworth, E. Y. *et al.* (1996). The Anglesey lakes, Wales, UK: changes in trophic status of three standing waters as inferred from diatom transfer functions and their implications for conservation. *Aquatic Conservation: Marine and Freshwater Ecosystems*, **6**, 81–92.

Cheffings, C. M. & Farrell, L. (eds.) (2005). *The Vascular Plant Red Data List for Great Britain. Species Status No.* 7. Peterborough: Joint Nature Conservation Committee.

Davies, C., Shelley, J., Harding, P. *et al.* (2004). *Freshwater Fish in Britain. The Species and their Distribution.* Colchester: Harley Books.

Department of the Environment (1994). *Biodiversity: The UK Action Plan.* London: Her Majesty's Stationery Office.

Dinerstein, E., Olson, D. M., Graham, D. J. *et al.* (1995). *A Conservation Assessment of the Terrestrial Ecoregions of Latin America and the Caribbean.* Washington, DC: The World Bank.

Duigan, C., Kovach, W. & Palmer, M. (2006). *Vegetation Communities of British Lakes: A Revised Classification.* Peterborough: Joint Nature Conservation Committee.

Duigan, C., Kovach, W. & Palmer, M. (2007). Vegetation communities of British lakes: a revised classification scheme for conservation. *Aquatic Conservation: Marine and Freshwater Ecosystems*, **17**, 147–73.

Duigan, C. A., Allott, T. E. H., Bennion, H. *et al.* (1996). The Anglesey lakes, Wales, UK – a conservation resource. *Aquatic Conservation: Marine and Freshwater Ecosystems*, **6**, 31–53.

Duker, L. (2001). *Report Series 2: Biodiversity Conservation of the World's Lakes: A Preliminary Framework for Identifying Priorities.* Maryland: LakeNet.

Earnst, S. L., Platte, R. & Bond, L (in review). A GIS-based model for yellow-billed loon habitat relationships in northern Alaska. *Hydrobiologia.*

England Field Unit (1990). *Handbook for Phase 1 Habitat Survey. A Technique for Environmental Audit.* Peterborough: Nature Conservancy Council.

Environment and Heritage Service (1999). *Guidelines for Selection of Biological ASSIs in Northern Ireland.* Belfast: Environment and Heritage Service.

Flower, R. J., Juggins, S. & Batterbee, R. W. (1997). Matching diatom assemblages in lake sediment cores and modern surface sediment samples: the implications for lake conservation and restoration with special reference to acidified systems. *Hydrobiologia*, **344**, 27–40.

Foster, G. N., Foster, A. P., Eyre, M. D. & Bilton, D. T. (1990). Classification of water beetle assemblages in arable fenland and ranking of sites in relation to conservation value. *Freshwater Biology*, **22**, 343–454.

Heegaard, E., Birks, H. H., Gibson, C., Smith, S.J. & Wolfe-Murphy, S. (2001). Species-environment relationships of aquatic macrophytes in Northern Ireland. *Aquatic Botany*, **70**, 175–223.

Heyer, W. R., Donnelly, M. A., McDiarmid, R. W., Hayek, I. C. & Foster, M. S. (1994). *Measuring and Monitoring Biological Diversity: Standard Methods for Amphibians.* Washington, DC: Smithsonian Institution Press.

Jefferies, D. J. (ed.) (2003). *The Water Vole and Mink Survey of Britain 1996–1998 with a History of the Long Term Changes in the Status of Both Species and their Causes.* Ledbury: The Vincent Wildlife Trust.

King, R. J. & Brackney, A. W. (1997). *Aerial Breeding Pair Surveys of the Arctic Coastal Plain of Alaska, 1996.* Unpublished Report. US Fish and Wildlife Service, Fairbanks, Alaska.

Langdon, R., Andrews, J., Cox, K. *et al.* (1998). *A Classification of the Aquatic Communities of Vermont.* Vermont: Aquatic Classification Workgroup for The Nature Conservancy and the Vermont Biodiversity Project.

Lyons, J. (1996). Patterns in the species composition of fish assemblages among Wisconsin streams. *Environmental Biology of Fishes*, **45**, 329–41.

Merritt, R., Moore, N. W. & Eversham, B. C. (1996). *Atlas of the Dragonflies of Britain and Ireland.* ITE Research Publication No. 9. London: Her Majesty's Stationery Office.

Moss, B., Johnes, P. & Phillips, G. (1996). The monitoring of ecological quality and the classification of standing waters in temperate regions: a review and proposal based on a worked scheme for British waters. *Biological Reviews*, **71**, 301–39.

Moss, B., Stephen, D., Alvarez, C. *et al.* (2003). The determination of ecological quality in shallow lakes – a tested expert system (ECOFRAME) for implementation of the European Water Framework Directive. *Aquatic Conservation: Marine and Freshwater Ecosystems*, **13**, 507–49.

Murphy, S. C. & Smith, D. W. (2001). Documenting trends in Yellowstone's beaver population: a comparison of aerial and ground surveys in the Yellowstone Lake Basin. In *Yellowstone Lake: Hotbed of Chaos or Reservoir of Resilience? Proceedings of the 6th Biennial Conference on the Greater Yellowstone Ecosystem. October 8–10, 2001, Mammoth Hot Springs Hotel, Yellowstone National Park*, eds. R. J. Anderson & D. Harmon. Yellowstone National Park, WY, and

Hancock, MI: Yellowstone Center for Resources and The George Wright Society.

Nature Conservancy Council (1989). *Guidelines for Selection of Biological SSSIs*. Peterborough: Nature Conservancy Council.

Noss, R. (1991). *Protecting Habitats and Biological Diversity. Part 1, Guidelines for Regional Reserve Systems*. New York: National Audubon Society.

Noss, R. F. & Cooperider, A. Y. (1994). *Saving Nature's Legacy: Protecting and Restoring Biodiversity*. Washington, DC: Defenders of Wildlife and Island Press.

Organization for Economic Co-operation and Development (1982). *Eutrophication of Waters. Monitoring, Assessment and Control*. Paris: OECD.

Palmer, M. (1973). A survey of the animal community of the main pond at Castor Hanglands National Nature Reserve, near Peterborough. *Freshwater Biology*, **3**, 397–401.

Palmer, M. & Smart, J. (2001). *Guidelines for the Selection of Important Plant Areas in Europe*. London: Plantlife.

Palmer, M. A. (In prep.). *LACON: Lake Evaluation for Conservation. Version 1 Manual*. Edinburgh: Scottish Natural Heritage.

Palmer, M. A., Bell, S. A. & Butterfield, I. (1992). A botanical classification of standing waters in Britain: applications for conservation and monitoring. *Aquatic Conservation: Marine and Freshwater Ecosystems*, **2**, 125–43.

Parsons, J. (2001). *Aquatic Plant Sampling Protocols*. Publication No. 01-03-017. Olympia, Washington: Environmental Assessment Program.

Poiani, K. A., Richter, B. D., Anderson, M. G. & Richter, H. E. (2000). Biodiversity conservation at multiple scales. *BioScience*, **50**, 133–46.

Preston, C. D. & Croft, J. M. (1997). *Aquatic Plants in Britain and Ireland*. Colchester: Harley Books.

Rahel, F. J. & Hubert, W. A. (1991). Fish assemblages and habitat gradients in a Rocky Mountain-Great Plains stream: biotic zonation and additive patterns of community change. *Transactions of the American Fisheries Society*, **120**, 319–32.

Ratcliffe, D. R. (ed.) (1977). *A Nature Conservation Review*. Cambridge: Cambridge University Press.

Regional Ecosystem Office, Regional Interagency Executive Committee (1995). *Ecosystem Analysis at The Watershed Scale: The Federal Guide for Watershed Analysis*. Sections I and II, Version 2.2.

Ricciardi, A. & Rasmussen, J. B. (1999). Extinction rates of North American freshwater fauna. *Conservation Biology*, **13**, 1220–2.

Rodwell, J. S. (ed.) (1995). *British Plant Communities. Volume 4. Aquatic Communities, Swamps and Tall-herb Fens*. Cambridge: Cambridge University Press.

Rowan, J. S., Soutar, I., Bragg, O. M., Carwardine, J. & Cutler, M. E. J. (2006a). Development of a technique for lake habitat survey (LHS): Phase 2. Edinburgh: Scotland and Northern Ireland Forum for Environmental Research (SNIFFER) (www.sniffer.org.uk/exe/download.asp?sniffer_outputs/WFD42 _1.pdf).

Rowan, J. S., Carwardine, J., Duck, R. W. *et al.* (2006b). Development of a technique for Lake Habitat Survey (LHS) with applications for the European Union Water Framework Directive. *Aquatic Conservation: Marine and Freshwater Ecosystems*, **16**, 637–57.

Stewart, N. F. (2004). *Important Stonewort Areas of the United Kingdom*. Salisbury: Plantlife International.

Strachan, R., Birks, J. D. S., Chanin, P. R. F. & Jefferies, D. J. (1990). *Otter survey of England 1984–1986*. Peterborough: Nature Conservancy Council.

The Nature Conservancy (1999). *Freshwater Initiative Workshop Proceedings: Evaluating Ecological Integrity at Freshwater Sites*. Virginia: Nature Conservancy.

Thoma, R. F. & McNight, C. (1995). Lake Erie biological criteria and habitat evaluation project. In *Methods of modifying habitat to benefit the Great Lakes ecosystem*, eds. J. R. M. Kelso & J. H. Hartig. CISTI (Canada Institute for Scientific and Technical Information). Occasional Paper No. 1, pp.191–196.

US Environmental Protection Agency (1999). *The Watershed Information Network*. Washington, DC: USEPA.

US Environmental Protection Agency (2000). *National Water Quality Inventory:2000 Report (EPA-841-R-02-001)*. Washington, DC: Environmental Protection Agency.

Usher, M. B. & Balharry, D. (1996). *Biogeographical Zonation of Scotland*. Edinburgh: Scottish Natural Heritage.

Williams, P., Biggs, J., Whitfield, M. *et al.* (1998). *Biological Techniques of Still Water Quality Assessment. Phase 2. Method Development*. Environment Agency R & D Technical Report E56. Bristol: Environment Agency.

Wolfe-Murphy, S. A., Lawrie, E. A., Smith, S. J. & Gibson, C. E. (1991). *Survey Methodologies: Data Collection Techniques*. A report by the Northern Ireland Lakes Survey. Belfast: Department of the Environment (Northern Ireland).

Wolfe-Murphy, S. A., Lawrie, E. A., Smith, S. J. & Gibson, C. E. (1992). *The Northern Ireland Lake Survey: Part 3. Lake Classification Based on Aquatic Macrophytes*. Belfast: Department of the Environment and Queen's University Belfast.

9 · *System Aqua – a Swedish system for assessing nature conservation values of fresh waters*

EVA WILLÉN

Introduction

The Nordic countries Sweden, Finland and Norway have a plethora of lakes, rivers and streams of various sizes. There are approximately 300 000 lakes larger than 1 ha in the three countries together, and several hundred thousand tarns, wetland-lakes and other small water bodies less than 1 ha. In Sweden, inland waters occupy approximately 10% of the land area (Figure 9.1). In Finland and Norway the corresponding figures are 20% and 5% (Seppälä, 2005). Lakes and rivers have long played a significant role in these countries as a means of transportation, as sources for fishing and hunting, as sources of energy, supplies of drinking water, etc. Already in the Middle Ages rivers were used for log-driving, and in small streams water power was utilized to operate water wheels at sawmills and flour-mills. These encroachments now have a considerable impact on fish movements.

Today the main parts of the larger watercourses are regulated to generate electric power, and the few rivers which still have mighty unrestrained waterfalls are protected under the Nature Conservation Act (in Sweden in 1964, in Norway in 1970) or in other ways excluded from exploitation. Many lakes, especially in lowland areas, suffer from land drainage and water level lowering to gain arable land to feed a growing population during an especially expansive period of the nineteenth century. Close to densely populated centres and in regions with fertile soils many lakes are nutrient-enriched. Another problem, counted in tens of thousands of affected lakes, is acidification caused by a high deposition of anthropogenically produced acid substances in combination with a prevalence of soils with a weak buffering capacity. The southern parts of Norway and Sweden are regions that are especially affected (Skjelkvåle

Assessing the Conservation Value of Fresh Waters, ed. Philip J. Boon and Catherine M. Pringle. Published by Cambridge University Press.

Figure 9.1. Distribution of lakes in Sweden. (See colour plate)

et al., 2001). In spite of the intense use of some of the waters and a regional exceedance of buffering capacity against acidifying matters, substantial parts of the lakes and watercourses are mainly unimpaired particularly in the northern half of the countries where the population pressure is low.

The richness of water bodies may have contributed to a certain ignorance concerning their protection compared with the early conservation interest devoted to natural or semi-natural terrestrial areas. Rather few waters have been protected with limnological values as a primary reason, although national parks and nature reserves cover considerable areas where many lakes and watercourses are situated. In the late 1950s, freshwater conservation was discussed internationally by a group of well-reputed limnologists resulting in a compilation of especially valuable lakes and watercourses mainly selected by criteria such as naturalness, research interest, and value for education and training. The selection was entirely based on expert opinion. Fifty water bodies were proposed in the three Nordic countries, and comprised the first formal attempt to grasp a broad international approach to water conservation (Luther & Rzoska, 1971).

Criteria used for nature conservation evaluation in the Nordic countries

In the early phase of selecting regions for conservation the aim was to protect Nature from Man. Wilderness, size and distance from areas with a high population density were then important criteria. Rarity was another significant criterion for conserving sites that had some unique feature such as a special plant or animal or some geological structure. In an investigation of early criteria used in Sweden, especially directed towards those used for the conservation of lakes, recreational value was particularly common, followed by botanical and ornithological values (Götmark & Nilsson, 1992). For rivers, geological values such as remarkable and representative formations were also prevalent. Since the 1960s, Norway has focused on a special plan for protecting certain watercourses from being exploited for hydroelectric power production. The plan, developed in three phases, now comprises 341 rivers and their catchments, and covers one-third of the total area of the country (Eie *et al.*, 1996). Expert judgement has been intensively used during the selection procedure based on criteria such as representativity, size, recreational value, cultural value and value for wild animals and fisheries. In the Norwegian protection plan those watercourses situated close to population centres, and therefore especially used and valued as recreational environments, were given priority – a case of protecting rivers for human use.

Criteria for assessing the conservation value of national parks and other protected areas have been established by IUCN and other international organizations (UNESCO, 1974). These criteria were used by the Swedish Environmental Protection Agency (SEPA) in the 1970s for selecting 600 lakes and watercourses as being nationally important to protect, but with the proviso of allowing a certain degree of human use. The following qualities were deemed to be particularly important:

- areas that are representative examples of landscape features;
- areas with a high degree of naturalness;
- areas with rare landscape characters and/or with endangered, vulnerable or threatened species;
- areas with high biodiversity;
- areas of a unique character.

In addition, a number of supporting criteria were suggested such as size, diversity of ecosystems, diversity of habitats or taxa, continuity, important areas for understanding landscape development, ecological functioning, vulnerability and the likelihood of maintaining the natural values (cf. Wiederholm, 1997). A 3-degree scale was suggested for rating the criteria, but a great deal of subjectivity was still unavoidable and different users produced different results.

There is no doubt that the criteria used for nature conservation evaluation in all Nordic countries have been qualitative to a large extent, implying that statements and expert judgements were made descriptively more than by some consistent ranking procedure. Although many assessments were carried out by professional biologists and geologists, they were also based on political considerations and opinions at the time of selection, which might later turn out to be irrelevant. In all Nordic countries, national guidelines have been established for assessing water quality, although at a relatively late stage of the twentieth century – guidelines which now are under revision for adaptation into the EC Water Framework Directive (WFD; see Chapter 2). These guidelines are aimed more at water quality assessment than for the purpose of conservation.

In Sweden as well as in Norway there have been further attempts to formalize a conservation evaluation procedure by rating water bodies against a set of criteria but still using a high degree of personal judgement (Berntell *et al.*, 1984, NPCA, 1989). In the mid-1990s the lack of an official and standardized instrument for catchment and site-specific characterization provided the incentive for a new approach to be used in country-wide assessments of lakes and watercourses in Sweden. The new tool, inspired by several international contributions (Ratcliffe, 1977; Rabe & Savage, 1979; O´Keeffe *et al.*, 1987; Ten Brink *et al.*, 1991; Dynesius & Nilsson, 1994) and especially the SERCON system developed in the UK (see Chapter 7; Boon *et al.*, 1994, 1997) was entitled System Aqua (Willén *et al.*, 1997). The SERCON system formed a special basis for discussing relevant indicators for evaluation and characterization but with adaptation to the geographical characteristics and the diversity of ecosystems in Sweden. SERCON also helped in providing a background to the scoring technique in System Aqua. This kind of formalized tool is not yet established in the other Nordic countries.

The System Aqua approach

System Aqua has been designed to meet the growing need for an objective, reproducible and comprehensive tool for assessing the conservation value of lakes and watercourses. A primary version internationally published in 1997 focused exclusively on characterization of biodiversity, interpreted as variations among organisms, habitats and ecosystems (Willén *et al.*, 1997). Based on large-scale regional tests, System Aqua was eventually developed further by including the views of the many regional experts working in Swedish county councils, and the focus was broadened to more multi-dimensional aspects of nature conservation (SEPA, 2001).

The general advantages of System Aqua are:

- Standard methods are proposed for field work, characterization and ecological assessments. The system thus allows uniform evaluation measures of nature conservation inventories.
- Data handling and characterization of Swedish lakes and watercourses and their catchments are standardized and can easily be used in conjunction with regional plans. Comparisons of waters in relation to size, stream order and other optional characters can be made from all parts of the country.
- Rivers and lakes deserving special consideration regarding their conservation value or their need for restoration are easily assessed.
- The use of scores, summary criterion indices reflecting assessed indicators and a final translation of the criteria profile for assessment of nature conservation value give comprehensible and easily accessible information to experts and also to non-experts.

- Continual adaptation is possible in order to meet the demands of the WFD and Natura 2000 (the network of conservation areas established under the EC Habitats Directive).

System Aqua has been constructed for use in several ways: first, to evaluate separate catchments by the use of maps, statistics and other printed sources; second, to evaluate a lake or watercourse nested in its catchment; and third, to concentrate on an individual lake or river reach. In Tier I an estimate can be obtained of the naturalness in a larger catchment (up to 300 km^2) and this forms the basis for further evaluations at a more detailed scale. Tier II gives the most complete evaluation since all criteria and indicators in the system can be used. Tier III has similarities with a traditional level of working with conservation evaluation of single water bodies and less consideration of the neighbouring land. Tiers II and III require a thorough investigation of the water body and its immediate surroundings, and for that purpose a standardized technique of habitat classification has been developed (Halldén *et al.*, 1997; Jacobson & Liliegren, 2000).

System Aqua has undergone several revisions, so that the present version now has a structure comprising four separate parts: one of characterization, another of evaluation, a third for describing special features and a fourth part for comprehensive assessment of natural value which provides a tool for potential decisions on conservation and management. The characterization part offers a uniform template for computerizing factual information, meeting the demands of many central and regional authorities. There are obvious advantages in collecting and computerizing information in a uniform and standardized way, such as the ability to sort the waters according to size, stream order, ecoregion or other features of interest. The characterization stage includes an assessment of structural diversity but no scoring is made. In the evaluation part, two of the criteria are used: naturalness and rarity. A third criterion – species richness – has a special status in System Aqua; it usually belongs to the characterization part except where water bodies have similar scores for naturalness and rarity. In these cases, the species richness score may be used to distinguish between two or more water bodies that otherwise have similar conservation value. The criteria are assessed by 3–7 indicators with the smallest number used in catchments and the largest for evaluating the naturalness of watercourses. Special features used in System Aqua are those which may be important but less amenable to scoring, such as amenity value, recreational use and uniqueness. These four parts of System Aqua – characterization, evaluation, special features and assessment of natural value – are the main pieces needed for identifying waters with particular conservation value or those with potential for mitigation or restoration.

Structure of the system

Identification and characterization

The main objective of the identification part is to provide basic geographical and morphological data in a consistent way. Structural diversity was used as a scored criterion in earlier drafts of System Aqua (Willén *et al.*, 1997) but is no longer scored in the final version because of its lack of an unambiguous relationship with high biodiversity and/or high natural values. For example, in several parts of Sweden the many inland areas have limited altitudinal range, where the lakes are mainly humic with low biodiversity but still with a high degree of naturalness. Conversely, in many of the alpine and sub-alpine lakes the structural and physical diversity is high yet biodiversity is extremely low, and catchment characteristics for several other attributes are uniform.

The following attributes are considered for characterizing structural diversity: number of land-use categories in the catchment and close to the lake/watercourse (at a distance of 30 m from the water´s edge); lake area as a proportion of catchment area; character of streams and fluvial features in a watercourse and shoreline development in a lake (deviation from a circular lake of the same area); topographic relief in a catchment; number of substrates and vegetation types in the water. A maximum of six vegetation types can be selected, i.e. emergent, floating-leaved, submerged with unbroken leaves, submerged with fine-branched leaves, submerged rosette plants and aquatic mosses.

Evaluation

The evaluation part concentrates on the criteria of naturalness and rarity. Naturalness is here defined either as a system lacking impact from human activities, or with interference sanctioned by long usage. A continuous gradient from natural/near-natural conditions to a completely non-natural state is considered when assessing ecological impact. Intact or almost intact ecosystems and their different subsystems should maintain long-term functioning within normal variations. Changes in intact systems depend on natural dynamics. Naturalness is affected by the presence of alien species, physical encroachment and various pollutants, whether produced locally or transported from other regions. Such perturbations cause changes in biodiversity, community structure and the entire ecosystem. The degree of naturalness in the flora and fauna of lakes and watercourses, including their adjacent environments, is an important indication of biological diversity and an important natural value. The chosen indicators of naturalness for catchments, lakes and watercourses are listed in Table 9.1.

Within the criterion 'rarity', red-listed species occurring in the water body, at the shore or those dependent on the specific watercourse or lake for survival, are evaluated irrespective of abundance. An especially large population of a rare species may be noted among 'special features'. In lakes, only species within the lake or in the shore area are evaluated, but in streams and rivers species are evaluated that occur both along the main channel and in tributaries (if the latter are not specifically assessed in their own right). Threatened species using the water environment temporarily as a halting place or migration route are evaluated also. The ecological function of a special water body in this respect may be important for the population in the region or even in the country as a whole. The five red-listed categories – critically endangered (CR), endangered (EN), vulnerable (VU), data deficient (DD) and near threatened (NT) – are scored under this criterion (Table 9.2), for the following biotic groups: plants (lichens, mosses, charophytes, vascular plants and algae); macro-invertebrates in the littoral and the profundal zones, crayfish, fish, amphibians, birds and mammals. Although there is a general awareness that the number of red-listed species recorded depends on survey effort, the use of the rarity criterion is prominent in nature conservation. It is hoped that the high scoring of some of the red-listed categories will stimulate authorities to support further inventories in waters where red-listed species may be expected. A basic requirement for evaluating rarity in System Aqua is to have data from at least one occasion of macroinvertebrate sampling and one of test fishing during the last 10 years. These two biotic groups are the most extensively investigated in Sweden.

Although species richness is mainly a criterion for characterization, where there are similar scores for naturalness and rarity species richness may be decisive for a final assessment. It is recommended that data ≤10 years old are used and for the sake of comparability certain conditions have to be fulfilled concerning sampling season, sampling frequency and sampling methods. Macrophytes (submerged and floating forms), littoral/profundal macro-invertebrates, fish and phytoplankton are the preferred indicators because standard methods for sampling are available and the organisms are frequently used for water quality assessment all over the country. Emergent macrophytes are not included in the assessment owing

Table 9.1 *Indicators evaluated under the criterion 'naturalness' in System Aqua*

Catchment	Lake/watercourse
Physical encroachment Degree of fragmentation (0 to >75%) of main channel by dams or artificial migration obstacles to salmon and trout*	Long-lasting encroachment Extent of changes in the river profile Extent of lake littoral zone (changed by raising or lowering water level, dredging, changes of outlets, etc. (0 to >75%)
Chemical impact Deviations from reference values outlined in Swedish Assessment Criteria (1999) concerning alkalinity and total phosphorus	Flood control of main channel Extent of regulation of water flow (0 to temporary dryness) Extent of legally documented damming in lake amplitude (0 to >3 m).
Land use Extent of severely impacted land-use types: clear-cut land, arable land, urban development	Land use in nearby shore area/riparian zone (0 – 30 m) Extent of shore length with severely degraded land-use types (in a range from <10 to >90%).
	Water quality Extent of deviation from reference values of minimum two variables: alkalinity/acidity, total phosphorus; metals are evaluated where appropriate (Swedish Environmental Quality Criteria 2000)
	Alien species Extent of impact on the indigenous flora and fauna
	Changes of flora and fauna Deviations from reference conditions (minimum variables for evaluation are fish and macro-invertebrates)
	Fragmentation Extent of dams or artificial migration obstacles for salmon and trout (0 to >75%) in rivers

*Degree of fragmentation expressed as the longest main channel without dams in relation to the total main channel. Main channel is the most water-rich channel section (Dynesius & Nilsson, 1994).
Source: SEPA (2001).

Table 9.2 *Lists of rankings given to the highest red-list categories found in an assessed lake or watercourse*

No. of critically endangered (CR) species	No. of endangered (EN) species	No. of data deficient (DD) species	No. of vulnerable (VU) species	No. of near threatened (NT) species	Score
≥1					5
	≥3				5
		≥5			5
	2				4.5
		4			4.5
	1				4
		3			4
			≥5		4
		2			3.5
			4		3.5
		1			3
			3		3
			2		2.5
			1		2
				≥5	2
				4	1.75
				3	1.5
				2	1.25
				1	1

Note that the score 5 can be reached by finding only one critically endangered (CR) species or by a combination of endangered (EN) and data-deficient (DD) species.

to their large variation in natural/near natural waters between northern and southern Sweden; in addition, the submerged and floating forms are especially used for water quality assessment in the country (Andersson & Willén, 1999; SEPA, 2000). In future, the species richness criterion may be expanded to include other indicators (e.g. breeding birds), but at present standardized and comparable surveys of those are limited.

Special features

'Special features' include those traits which are hard to score because they include a large amount of subjectivity. The features may be weighted either as positive or negative in the overall conservation assessment. The feature 'uniqueness' is one of those listed, and this may be highly decisive if it refers to some sole or rare representative of its kind at a regional or country level. Examples of other positive features include important areas for breeding birds, research value, relationship to valuable cultural structures, raw water resource, importance for commercial or recreational fishing, areas of important ecological functioning (e.g. spawning areas for rare fish), occurrence of genetically unique races of fish, etc. Features considered as negative in the overall assessment are the occurrence of aquaculture, damage from tourist activities in nearby areas, and other large encroachments which may affect the quality of the water or the immediate corridor (gravel pits, plans for holiday settlements, expanded road construction plans, etc.).

Scoring indicators and criteria

Indicators of the criteria naturalness, rarity and species richness are scored in a range from 0 to 5 where 5 is allocated to the most favourable situation from a nature conservation point of view, and 0 indicates a highly disturbed environment. The total score of each criterion also has the same range, while the integrated scoring of the natural value in a catchment, a lake or a watercourse is presented in a descriptive scale from very high through high, moderate, low to very low, i.e. a 5-degree scale (cf. Table 9.5).

The summary score for the criterion naturalness is a mean value of the scores for the indicators tested, while the summary score for rarity is a weighted value where species belonging to the class of CR-species are given the highest score of 5. The presence of one species in the EN-category scores 4, but the score increases to 4.5 and 5 with the occurrence of two and three species, respectively. Species categorized as DD and VU are downweighted by a factor of 0.5, and those classified as NT by 0.25. To facilitate calculations and to avoid misinterpretations, rankings are given for the species placed in the highest category in an assessed water body, which, for example, could be the category of EN species (Table 9.2). The total score after evaluation of several red-listed species follows the formula:

Total score = score of the highest ranked category
(Table 9.2) + $(0.5 \times X_3) + (0.5 \times X_4) + (0.25 \times X_5)$, where
X_3 = number of species within the category DD
X_4 = number of species within the category VU
X_5 = number of species within the category NT

The weighted combined rarity score in a lake or stream may exceed 5, but 5 is the maximum score allocated. The red-listed species will nevertheless

Table 9.3 *Scoring of species richness of submerged and floating-leaved macrophytes and indigenous fish in Swedish lakes*

Score	No. of macrophyte species	No. of indigenous fish species
5	>18	≥10
4	15–18	6–9
3	10–14	3–5
2	5–9	2
1	1–4	1
0	0	0 (treated with rotenone)

Source: SEPA (2001).

be registered which may give a decisive weight to water bodies with otherwise similar overall scores.

The assessments under the criterion rarity may change in accordance with how the ranking of the various threatened categories are interpreted, and as the red lists are updated every 5 years the latest editions always have to be checked. Even though species richness is normally used just for characterization, the indicators are all scored so that the criterion can be used when two or more waters have the same summary score. Table 9.3 shows the score bands for macrophytes and fish in lakes. For macroinvertebrates different ranges are given for littoral, sublittoral and profundal samples and for two different levels of analysis. Two ranges are also given for phytoplankton: one related to genus and the other to species level. A prescribed species list is used for characterization and only species recorded using standard sampling methods and analyses are used. The combined score is a mean value of all indicators, but principally macroinvertebrates and fish.

Many lakes in Sweden (mainly in the alpine or sub-alpine regions) are naturally fishless; these are not evaluated in System Aqua.

Interpretations of System Aqua outputs

The information used in System Aqua is stored in a database where it is possible to sort the material for many different purposes. For comparisons of catchments or individual lakes or rivers it is most appropriate to compare those of a similar size and topographic range, similar stream order and ecoregion. It is also possible to select and compare waters with similar structural diversity in the catchment. In Sweden, this type of integrated

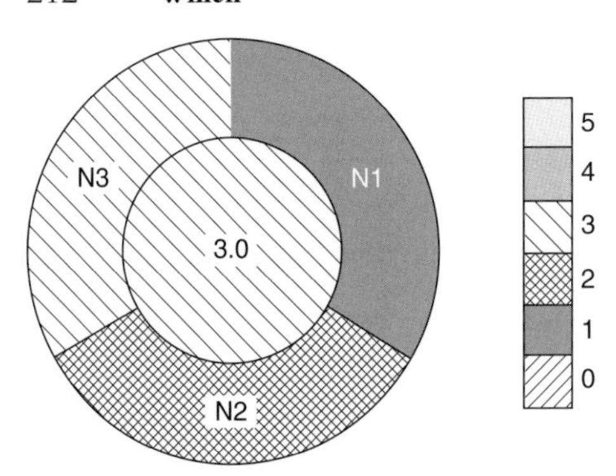

Figure 9.2. Scores for the criterion naturalness in a catchment. The mean score is given in the centre and the scores of the three indicators, physical encroachment (N1), chemical impact (N2) and land-use types (N3), are interpreted from the patterned scale.

and universal database for inland waters is unique, and it will facilitate the work of the county councils by highlighting local, regional and national variations.

An example of the scoring given for naturalness in a catchment is shown in Figure 9.2, where the records of the three indicators are illustrated by different patterns and in the centre the mean score of the criterion is given. If using colours, a range of tints from blue through green and yellow to red will denote the score of naturalness from natural water bodies to those that are highly degraded. As System Aqua can also make regional evaluations, which deviate slightly from the national one, the final evaluation may be separated on a national and a regional scale. The outcome of the catchment evaluation illustrated in Figure 9.2 is a moderate degree of naturalness. Similar types of pie chart are used to present the results from evaluations of naturalness, rarity and species richness in different water bodies (Figure 9.3). The graphs are readily interpretable, showing clearly the scores given for each indicator within the two criteria. White sectors reveal indicators where data are missing. If the minimum requirements set for evaluation of some indicator are not fulfilled it is still possible to get a summary score, but indicate that by a hatched pattern for the indicator in question. This is illustrated in Figure 9.3 where the inventory of the A2 indicator did not follow the recommended standard. In a case like that the total score would be shown also by a hatched pattern. It is hoped that such an outcome will stimulate regional or national authorities to make up the shortfall in data. Until then, the evaluation should be used with caution. As mean values and weighted values may result in a decimal number, a key for interpreting those values is included in the system (Table 9.4).

Table 9.4 *Transference of mean values to a criterion score for naturalness (N), rarity (R) and species richness (S)*

Mean value	Criterion score: N, R, S	Colour on map	Interpretation: Naturalness	Rarity	Species richness
≥ 4.6	5	Dark blue	Very high degree	Very high degree	Very species rich
≥ 3.8 to < 4.6	4	Light blue	High degree	High degree	Species rich
≥ 2.8 to < 3.8	3	Green	Moderate degree	Moderate degree	Moderately species rich
≥ 1.8 to < 2.8	2	Yellow	Low degree	Low degree	Moderately species poor
≥ 0.8 to < 1.8	1	Orange	Poor degree	Poor degree	Species poor
0.0 to < 0.8	0	Red	No naturalness	No known rarity	No occurrence of biota

Note: Examples relate to evaluation of a lake or a reach of a watercourse.

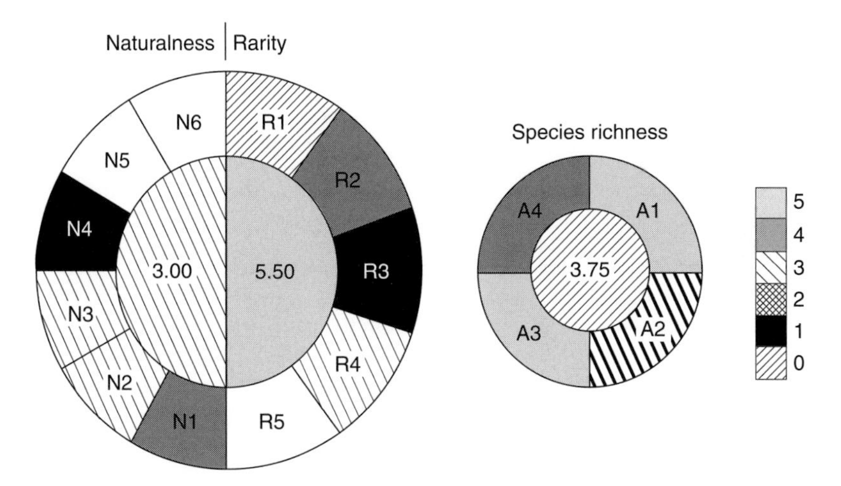

Figure 9.3. Scores for the criteria naturalness, rarity and species richness in a lake or watercourse. The mean score is given in the centre while scores of each indicator are interpreted from the patterned scale. A hatched pattern indicates that basic information required for an evaluation is not available, as shown in the centre of the species richness diagram.

An outline of the evaluation of a larger catchment and a number of sub-catchments is given in Figure 9.4, where the largest pie chart shows the outcome of naturalness in the whole catchment (A) including three stream orders, while C and B show evaluations of the first and second stream

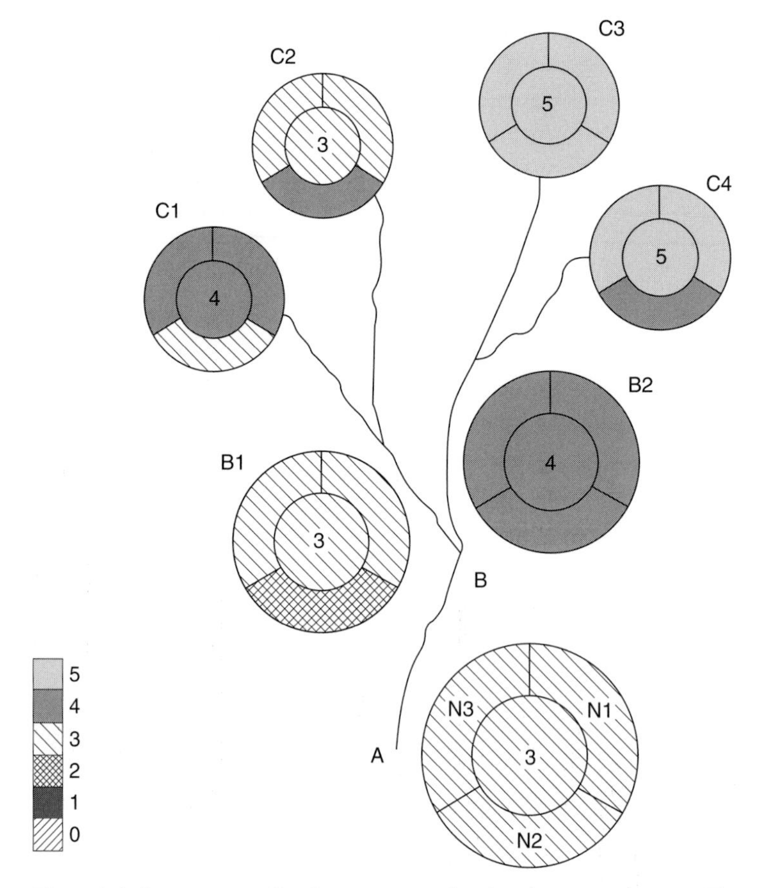

Figure 9.4. Assessments of various stream orders in a larger catchment. The criterion used is naturalness and the indicators are the same as in Figure 9.2. Indicator values and the mean criterion score are interpreted from the patterned scale.

order catchments. Such a comprehensive illustration helps the user to make decisions on where to carry out more detailed assessments to find waters potentially important for conservation.

Assessing natural value and the potential for nature conservation

A total, integrated score for natural value, based on the criterion summary scores, is produced as a final outcome by descriptions that range from 'very high' through 'high', 'moderate', 'low' to 'very low' values (Table 9.5). Naturalness is considered the most important attribute, but a high score for rarity may also justify a very high natural value for a moderately disturbed water body. A check of the scores for the indicators within naturalness

Table 9.5 *Final assessment of the natural value of a water body based on a total evaluation using System Aqua*

Naturalness	Rarity	Species Richness	Special features	Assessment of natural value
4–5	–	–	–	Very high
3	5	–	–	Very high
3	1–4	–	–	High
2	5	–	Unique or with very high structural diversity	High (measures to regain a very high value are desirable)
3	0	–	–	Moderate
0–2	4	4–5	–	Low (mitigation/restoration recommended if possible to increase naturalness)
0–2	0–3	0–3	–	Very low

Note: – indicates that the criterion score is not used for the total evaluation.

will reveal where mitigation might be made. Additional information under special features is another way of raising the conservation value, or sometimes decreasing it. When values for both naturalness and rarity reach a high score (i.e. 4–5 and 5, respectively), the interpretation is straightforward because then the water body is assigned a very high natural value. Although it is not always possible to provide statutory protection for all waters with high natural values, it is still important to bring those waters to public attention.

Future development

Further work to amplify System Aqua to meet the demands of the WFD has recently been initiated (Bergengren & Bergquist, 2004). The characterization part of the system has been extended to include a WFD typification of water bodies. Under the criterion naturalness and the indicator water quality, deviations from a natural state are assessed against a stricter definition of reference state than before, and lower classes are derived from that. An advantage is that WFD 'biological elements' are given priority, although physico-chemical and hydromorphological indicators will give additional information. The present combined total assessment of the natural value, and the potential for nature conservation in System Aqua, is ranked in a 5-degree scale similar to that recommended for scoring in the WFD. It is hoped that System Aqua will inspire other Nordic countries which have not yet devised a comprehensive scoring system for nature conservation evaluation. In addition, the European network of valuable habitats – Natura 2000 – protected under the EC Habitats Directive will need a method for characterizing and assessing conservation value in an understandable and reproducible way. For such purposes System Aqua is an appropriate tool for dealing with freshwater systems.

References

Andersson, B. & Willén, E. (1999). Lakes. In *Swedish Plant Geography*, eds. H. Rydin, P. Snoeijs & M. Diekmann. *Acta Phytogeographica Suecica*, **84**, 149–68.

Bergengren, J. & Bergquist, B. (2004). *System Aqua – Part 1. An Hierarchical Model for Characterization of Lakes and Watercourses* (in Swedish). County Council of Jönköping. Report 24.

Berntell, A., Wenblad, A., Henriksson, L., Nyman, H. & Oskarsson, H. (1984). *Criteria for Evaluation of Lakes for Conservation* (in Swedish). County Council of Älvsborg 1983: 3. Vänersborg.

Boon, P. J., Holmes, N. T. H., Maitland, P. S. & Rowell, T. A. (1994). A system for evaluating rivers for conservation (SERCON): an outline of the underlying principles. *Verhandlungen der Internationalen Verienigung für theoretische und angewandte Limnologie*, **25**, 1510–4.

Boon, P. J., Holmes, N. T. H., Maitland, P. S., Rowell, T. A. & Davies, J. (1997). A system for evaluating rivers for conservation (SERCON): development, structure and function. In *Freshwater Quality: Defining the Indefinable?* eds. P. J. Boon & D. L. Howell. Edinburgh: The Stationery Office, pp. 299–326.

Dynesius, M. & Nilsson, C. (1994). Fragmentation and flow regulation of river systems in the northern third of the world. *Science*, **266**, 753–62.

Eie, J. A., Faugli, P. E. & Aabel, J. (1996). *Rivers and Streams. Protection of Norwegian Watercourses* (in Norwegian). Oslo: Grondahl and Dreyers.

Götmark, F. & Nilsson, C. (1992). Criteria used for protection of natural areas in Sweden 1909–1986. *Conservation Biology*, **6**, 220–31.

Halldén, A., Liliegren, Y. & Lagerkvist, G. (1997). *Habitat Classification of Watercourses and their Environments* (in Swedish). County Council of Jönköping. Report 25.

Jacobson, C. & Liliegren, Y. (2000). *Habitat Classification of Watercourses and their Environments* (in Swedish). County Council of Jönköping. Report 24.

Luther, H. & Rzoska, J. (1971). *Project Aqua – A Source Book of Inland Waters Proposed for Conservation*. International Biological Programme (IBP), Handbook 21. Oxford: Blackwell Scientific Publications.

NPCA (1989). *Water Quality Criteria for Fresh Water* (in Norwegian). Oslo: Norwegian Pollution Control Authority.

O'Keeffe, J. H., Danilewitz, D. B. & Bradshaw, J. A. (1987). An expert system approach to the assessment of the conservation status of rivers. *Biological Conservation*, **40**, 69–84.

Rabe, F. W. & Savage, N. (1979). A methodology for the selection of aquatic natural areas. *Biological Conservation*, **15**, 291–300.

Ratcliffe, D. (ed.) (1977). *A Nature Conservation Review*. Cambridge: Cambridge University Press.

SEPA (Swedish Environmental Protection Agency) (2000). *Swedish Environmental Quality Criteria. Lakes and Watercourses*. Report 5050. Stockholm.

SEPA (2001). *System Aqua* (in Swedish). Report 5157. Stockholm.

Seppälä, M. (ed.) (2005). *The Physical Geography of Fennoscandia*. Oxford: Oxford University Press.

Skjelkvåle, B., Mannio, J., Wilander, A. & Andersen, T. (2001). *Recovery from Acidification of Lakes in Finland, Norway and Sweden 1990–1999. Hydrology and Earth System Sciences*, **5**, 327–37.

Ten Brink, B. J. E., Hosper, S. H. & Colijn, F. (1991). A quantitative method for the description and assessment of ecosystems: the AMOEBA-approach. *Marine Pollution Bulletin*, **23**, 265–70.

UNESCO (1974). *Task Force on Criteria and Guidelines for the Choice and Establishment of Biosphere Reserves*. Man and the Biosphere Report 22, Paris.

Wiederholm, T. (1997). Assessing the nature conservation value of fresh waters: a Scandinavian view. In *Freshwater Quality: Defining the Indefinable?* eds. P. J. Boon & D. L. Howell. Edinburgh: The Stationery Office, pp. 353–68.

Willén, E., Andersson, B. & Söderbäck, B. (1997). System Aqua: a biological tool for Swedish lakes and watercourses. In *Freshwater Quality: Defining the Indefinable?* eds. P. J. Boon & D. L. Howell. Edinburgh: The Stationery Office, pp. 327–33.

10 · *Evaluating Australian fresh waters for nature conservation*

JON NEVILL AND ANDREW BOULTON

General introduction

Australia is the driest inhabited continent, with its population of around 20 million people mostly concentrated on the well-watered eastern seaboard. Despite only 200 years of European occupation, water quality and native freshwater ecosystems have deteriorated seriously owing to over-extraction, flow regulation, pollution (especially salinization and eutrophication), urbanization and exotic pest invasion (reviews in Boulton & Brock, 1999; Schofield *et al.*, 2000; Arthington & Pusey, 2003). Although European settlement in Australia has been largely limited to the coastal cities (and nearby regions) of Sydney, Melbourne, Brisbane, Perth and Adelaide (Figure 10.1), inland waters have also been heavily exploited and regulated. Water resource development during the nineteenth century focused on water supply and waste disposal, while in the arid inland, successful agriculture and mining often relied on isolated feats of engineering to provide a reliable water resource (Evans, 2001). In this first century of European occupation, there was limited impact upon the landscape yet considerable development of legal and institutional arrangements relating to water supply (Smith, 1998) – many of which remained in place until the early 1990s.

Apart from temporary lapses during the First and Second World Wars and the Depression of the 1930s, Australian water resource development in the twentieth century escalated. Total dam capacity rose by orders of magnitude from 250 GL in 1900, to 8730 GL in 1940 and 87 260 GL in 1990, mostly as on-river storages in the Murray–Darling Basin (Crabb, 1997) and along the eastern seaboard. In the last 30 years, ample evidence of the environmental impacts of these spectacular engineering schemes has heralded awareness of the need to evaluate Australia's fresh waters for nature conservation. Attention has shifted from engineering issues of water quantity and supply to 'market liberalization' of water rights and ecological issues of environmental flows and water quality (ARMCANZ, 2000).

Assessing the Conservation Value of Fresh Waters, ed. Philip J. Boon and Catherine M. Pringle. Published by Cambridge University Press.

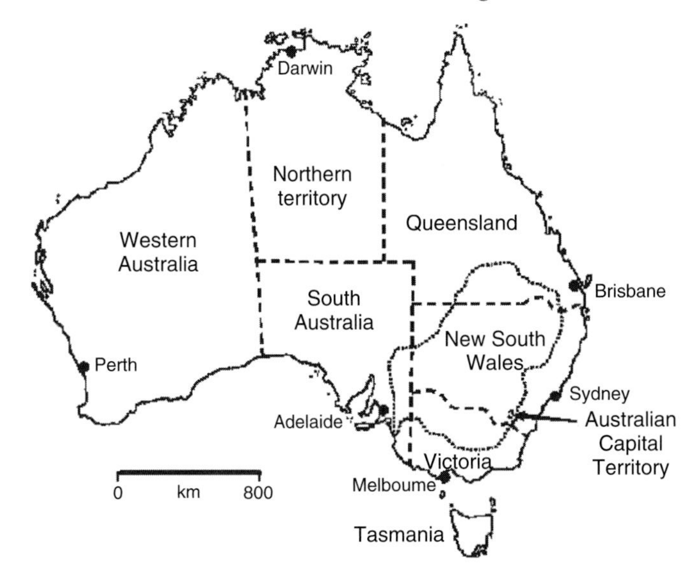

Figure 10.1. Australia's states, territories and capital cities mentioned in the text. The Murray–Darling Basin catchment is marked by the dotted line.

In this chapter, we describe the philosophical context of water resource management in Australia and how it is affected by the 'piecemeal delegation' of planning from state to local governments. The serious ecological threats that face freshwater ecosystems in Australia are exacerbated by the continued failure by state governments to apply effective controls on the cumulative effects of incremental water-related development (Finlayson *et al.*, 2008). We define 'threats' as processes likely to degrade ecosystem values, and 'values' as aspects of the ecosystem's condition that are valued by humans (in this case, conservation values because of the presence of rare species or natural habitat features, for example). 'Condition' identifies departure of an ecosystem from its natural state in terms of 'state variables' such as fish species diversity and 'functional variables' such as production to respiration ratios. Finally, 'importance' equates to degrees of value, and in the Australian context is applied at local, state, national and international scales.

Attention currently focuses on conducting wetland inventories across the states, with only limited scientific or political responses to broad-scale threats such as over-extraction or salinization. Although all Australian governments are committed to declaring protected areas (including representative examples of all major freshwater ecosystems), these commitments have yet to be effectively implemented (Nevill & Phillips, 2004).

Philosophical considerations

Understanding freshwater conservation in Australia requires recognition of the political structure of this Commonwealth country. Australia has a three-tiered governmental structure, with the first layer, the Australian (Federal or Commonwealth) government responsible for taxation, defence, economic regulation and international affairs, including decisions to support international treaties, agreements and programmes. Six states and two territories (Figure 10.1) form the second layer, and perform functions of health, education, law enforcement, social services and resource management – including the management of freshwater resources and ecosystems. Local (municipal) governments comprise the third tier, and are accountable for the delivery of land-use planning and some local health, welfare and environmental services.

The delegation of strategic land-use planning from state to local government levels invites 'piecemeal protection' and loss of strategic direction over localized development. Events of the last few years have provoked groups of respected scientists to seek to influence resource development debates. For instance, prompted by calls for engineering solutions to Australia's recent drought, the Wentworth Group released a 'blueprint' focused on national water resource management, especially water market and landscape-scale conservation issues (Wentworth Group of Concerned Scientists, 2002, 2003). At the national level, the Wentworth Group seems to have had only limited success in influencing policy development. For example, the National Water Initiative (Commonwealth of Australia, 2003), prepared by the Commonwealth government in consultation with the Council of Australian governments, appears to have ignored the Group's call for 'comprehensive water accounts' (a strategy aimed at the management of the cumulative effects of incremental development) and given only token recognition to the need to develop a national programme to protect rivers of high conservation value – another of the Group's key recommendations. Another collective has concentrated on protecting Australia's tropical rivers, many of which are still in relatively natural condition (Australian Tropical Rivers Group, 2004), but it is too early to judge this group's success in influencing key policy decisions by State governments.

Against this background of political controversy, several Australian scientists have tried to establish a conservation philosophy framework, borrowing heavily from methods used in the UK and the USA (see Chapters 7 and 8), with suitable refinements to acknowledge the country's highly variable natural climate (Dunn, 2000; Bennett *et al.*, 2002; Government of Victoria, 2002). Australia is fortunate in having a relatively small human population and still possessing many reasonably intact freshwater ecosystems, especially in tropical northern areas. However, even these northern fresh waters are not immune to threats of water-development schemes (Australian Tropical Rivers Group, 2004). There is still a need to convince politicians that protection is cheaper than rehabilitation, and that we must adopt national scales of protection for Australia's best freshwater ecosystems.

Protecting the best

International agreements

The Australian government has signed international agreements to protect natural areas, including aquatic ecosystems. In practice, owing to restrictions in the Australian constitution, protection is generally effected by state and territory jurisdictions. The international agreements most directly relevant here are the Convention on Wetlands 1971 (i.e. the Ramsar Convention) and the Convention on Biological Diversity 1992 (see Chapter 2). Under the Ramsar Convention, parties must nominate suitable sites as 'Wetlands of International Importance', manage them to maintain their ecological values, develop national systems of wetland inventories and wetland reserves and cooperate with other nations regarding wetlands and their resources (e.g. migratory birds). Australia was the first country to become a party to the Convention and the first to nominate a site to the Ramsar list. Theoretically, all states have programmes in place designed to meet commitments under the Ramsar Convention, particularly relating to the development of freshwater ecosystem inventories and the establishment of systems of reserves covering the full range of wetlands in the Ramsar definition of the term (including flowing and subterranean aquatic ecosystems). In practice, these programmes remain incomplete in all jurisdictions except the Australian Capital Territory (Nevill, 2007b) but work continues, particularly on ecosystem inventories.

The *Directory of Important Wetlands in Australia* (Environment Australia, 2001), compiled in response to commitments made under the Ramsar Convention, contains information on site value, but not condition. At this stage (October 2008), the Directory is not comprehensive in its approach or coverage as its listing concentrates on standing-water environments, although riverine and subterranean ecosystems are being added (Nevill,

2006). Acknowledgement of the importance of protecting aquatic subterranean and groundwater-dependent ecosystems (GDEs) in Australia is relatively recent (Boulton *et al.*, 2003) despite their significance in such an arid landscape where most surface water is ephemeral (Hatton & Evans, 1998).

The second major international agreement aimed at developing strategic protection frameworks is the Convention on Biological Diversity 1992, ratified by Australia in 1993, which requires (amongst other issues) that signatories identify, protect and monitor the health of major ecosystems. This Convention committed Australian governments to establish strategic systems of protected areas, and was reinforced in February 2004 when a revised programme of work on inland waters was adopted by the 7th Conference of the Parties to the Convention on Biological Diversity. The adopted measures include Goal 1.2: 'to establish and maintain comprehensive, adequate and representative systems of protected inland water ecosystems within the framework of integrated catchment/watershed/river-basin management' (Conference of the Parties, 2004).

To fulfil this obligation, Australia needs to classify rivers, lakes, ephemeral wetlands and subterranean ecosystems according to criteria which will allow *representative* types and values to be identified, and *representative* protected areas selected and managed. Inventories supporting such classifications are currently under development across the states and territories. Although development of a national inventory is being discussed (Nevill & Phillips, 2004; Kingsford *et al.*, 2005), no formal agreement has yet been reached amongst the jurisdictions to cooperate in this activity.

National agreements, strategies and protected area programmes

The establishment of systems of representative protected areas is identified as a commitment of all Australian governments in several key strategies (Commonwealth of Australia, 1992a,b, 1996). Consequently, all Australian jurisdictions recognize that a system of protected areas must represent ecosystem biodiversity, including freshwater biodiversity. However, development of terrestrial and marine protected areas has out-paced that of freshwater protected areas in Australia (Nevill & Phillips, 2004). Large terrestrial protected areas presumably protect aquatic ecosystems contained within them, but we need to know more of the catchment and subsurface linkages before accepting this assumption (Saunders *et al.*, 2002).

Tasmania and the ACT have high proportions of their land in protected areas. For example, 40% of the land area is protected in Tasmania, including a large World Heritage Area in the south west where several important rivers have been dammed to provide hydroelectric power. Such large protected areas do conserve many of the area's aquatic ecosystems, but for many other parts of Australia where land protection on this scale is not feasible, protection of 'linear' systems such as rivers is inadequate. Lowland rivers and their floodplains are a magnet for human settlement and agriculture, and there are no adequately protected lowland rivers in southern Australia (Stein, 2006). Instead, efforts focus on restoring elements of riparian zone vegetation, natural flow regime and water quality (Arthington & Pusey, 2003).

Responding to environmental threats

In Australia, freshwater ecosystems face serious threats from introduced species, extraction and regulation of water flow, channelization and desnagging, and catchment land-use changes that cause water pollution (e.g. salinization, eutrophication) or have direct impacts on aquatic species (Lake, 1995; Boulton & Brock, 1999; Nevill & Phillips, 2004). Most of these threats interact. For example, river regulation and changed flow regime in many of Australia's rivers have favoured invasion by exotic species such as carp, while construction of dams and weirs have prevented migration and recolonization by native fish (Koehn, 2001). Groundwater overdraft is pervasive and results in damage to GDEs, both aquatic and terrestrial (Evans, 2007; Nevill, 2007a). Little is known about human impacts on subterranean freshwater systems and GDEs, hampering their rehabilitation or protection (Boulton *et al.*, 2003; Boulton, 2005).

Australia's history of response to environmental threats to fresh waters is poor. Typically, the responses have focused more on treating the symptoms than addressing the causes, with predictably short-term 'band-aid' effects (Boulton & Brock, 1999) and generally tackling the issues at a small scale in a piecemeal fashion (Lake, 2005). Most evaluation of threats is done by professional consultants at the behest of regulatory agencies, usually when specific development proposals are planned that might risk damaging freshwater ecosystems. Inevitably, adequate long-term assessments are unlikely and seldom even contemplated.

Evaluating restoration potential

National assessments of river and wetland condition have identified sites or entire systems where severe degradation has occurred (National Land

and Water Resources Audit, 2001), prompting calls for restoration. Historically, largely guided by local government pressure and the piecemeal management approach described above, immense sums of money (over $A50 million annually, White *et al.*, 1999) were spent controlling erosion in rivers to protect assets such as valuable land and bridges by 'river restoration'. More recently, the wisdom of pouring money into restoring the most degraded areas has been questioned, and a more balanced approach has been suggested for setting priorities for rehabilitation and evaluating restoration potential in Australian fresh waters (Rutherfurd *et al.*, 2000, 2004). These priorities should be determined based on the values (such as natural biodiversity) that can be obtained for the time, money and effort available for the restoration activity. Rutherfurd *et al.* (1999) argued for investing more effort in protecting sites that remain in good condition rather than spending huge amounts to rehabilitate damaged areas. They also proposed a hierarchical approach to defining priorities from the national to the local scale and from large catchments down to sub-reaches.

Evaluating restoration potential goes beyond assessing the condition of the wetland; it must also consider the rarity of the asset in the region, the trajectory in condition (is it getting better or worse?), the ease with which the restoration can occur, and the degrading processes and how these can be treated (Rutherfurd *et al.*, 1999; Linke *et al.*, 2007). The first of these criteria could be extended to include aspects of biodiversity, importance of ecological functions and possession of rare or threatened communities – all values that could be reported for fresh waters in a national inventory. Assessment of the trajectory of change in the condition is more complex because in Australia, assessments of condition have been so infrequent that temporal changes are rarely known. Part of the need for this knowledge is to determine whether restoration efforts might hamper natural rates of recovery (Palmer *et al.*, 2005) or whether the effects of degrading processes could be prevented or ameliorated to stop further decline in condition.

One popular approach for assessing geomorphic river condition in Australia, the 'river styles' procedure (Brierley & Fryirs, 2000), considers the change from the expected condition, and has been used to predict recovery trajectories in some rivers (Brierley *et al.*, 2002). This should not be interpreted to mean that stages in restoration of a river reach should follow the same stages as its degradation, but such an approach assists practitioners and scientists understand some of the mechanisms of degrading processes (the fourth criterion above). With such an understanding, restoration is likely to be more successful and long-lasting than if only symptoms are treated and the degrading processes continue. As wetlands and rivers are interconnected systems, trajectories of recovery and degrading processes are also influenced by conditions in the catchment and along the river continuum, so no evaluation of restoration potential can be made successfully at a local scale alone (Lake *et al.*, 2007).

Evaluation of the ease of restoration includes the location of the wetland or river, the type of degrading processes, land tenure type (i.e. whether public, leasehold or freehold) as an indicator of the ease of establishing coherent management and the availability of local materials and labour. In this sense, evaluation of restoration potential must have a pragmatic aspect (Rutherfurd *et al.*, 1999) and tackle the seriously degraded sites later than less damaged ones. In reality, public opinion and local community pressure often overwhelm the logic of the approaches described above, and disproportionate amounts of restoration effort occur in urban streams or those in areas where influential landholders have been able to garner funds from local government.

Australia currently lacks an agreed national approach to evaluating restoration potential of its fresh waters. While the logic of protecting the best before restoring damaged systems is clear, the allocation of funding for this activity does not yet match demand (White *et al.*, 1999). Where restoration activities have been undertaken in Australian wetlands and rivers, evaluation of their success has been virtually absent, preventing managers and scientists from learning adaptively from opportunities provided by these large-scale 'field experiments' (Lake, 2001, 2007). Without this information on the success of restoration strategies, efforts to evaluate restoration potential are hampered because we are unable to judge the likelihood of a given strategy to succeed.

Methods for assessing conservation value of rivers and lakes

Scoring versus systematic approaches

Planning programmes need to identify (i) concordance of high conservation values with high condition as the most effective areas for proactive conservation management (particularly where threat is high), and (ii) concordance of high value with low ecological condition as potential priority rehabilitation areas, as already mentioned. Historically, reservation of wetland areas in Australia has been *ad hoc*, driven by local community pressures and opportunities provided by low-cost land. Systematic conservation planning (Margules & Pressey, 2000), while underpinning many recent terrestrial and marine reservations, has seldom been applied to freshwater conservation in Australia or overseas.

Scoring approaches are still being used, largely driven by the lack of an adequate national freshwater ecosystem database. Kingsford *et al.* (2005) and Nevill and Phillips (2004, Appendix 19) suggested the use of scoring techniques based on seven criteria:

Criterion 1: The ecosystem and its catchment are largely undisturbed by the direct influence of modern human activity.
Criterion 2: The ecosystem is a good representative example of its type or class within a bioregion or sub-bioregion.
Criterion 3: The ecosystem is the habitat of rare or threatened species or communities, or is the location of rare or threatened or significant geomorphic or geological feature(s), or contains one of only a few known habitats of an organism of unknown distribution.
Criterion 4: The ecosystem demonstrates unusual diversity and/or abundance of features, habitats, communities or species.
Criterion 5: The ecosystem provides evidence of the course or pattern of the evolution of Australia's landscape or biota.
Criterion 6: The ecosystem provides important resources for particular life-history stages of biota.
Criterion 7: The ecosystem performs important functions or services within the landscape (e.g., refugia, sustaining associated ecosystems).

Conservation priorities resting on opportunistic reservations will not produce a protected area network that comprehensively protects biodiversity (Pressey & Taffs, 2001). Systematic conservation planning has evolved in the past 25 years to provide a more rigorous, defensible and transparent basis for setting spatial conservation priorities. The objective is to design systems of protected areas that capture target amounts of biodiversity features for a minimal cost (Margules & Pressey, 2000) while promoting the persistence of fundamental ecosystem processes. The approach can be used to make spatially explicit decisions about a variety of conservation actions, including invasive species control and restoration of native ecosystems. Surprisingly, systematic conservation planning tools have only recently been applied to freshwater systems (Chadderton *et al.*, 2004; Higgins *et al.*, 2005; Nel *et al.*, 2007; Sowa *et al.*, 2007).

Spatially explicit index-based or scoring approaches have been commonly used to prioritize freshwater systems for conservation such as the establishment of Ramsar reserves, and were used in many broad-scale terrestrial assessments (e.g. global biodiversity hotspots based on species richness or rarity). Scoring approaches have the benefits of being explicit and consistent, usually combining several relevant considerations for conservation priority. However, they also have several important limitations (Smith & Theberge, 1987), including:

1. Combining rankings for criteria can be mathematically invalid and not meaningful (e.g. naturalness + species richness – threat = ?).
2. Outstanding scores for one or more criteria can be averaged out by low scores (a high score for fish should not be nullified by a low score for waterbirds).
3. Stopping rules for conservation action are often entirely arbitrary (how far down a list of priorities should planners go?).
4. It is usually not feasible to represent all conservation assets in a set of highest-scoring areas because scoring lists do not recognize complementarity (defined below).

Systematic conservation planning, developed in response to the limitations of scoring approaches (Kirkpatrick, 1983; Pressey, 2002), has several important advantages. The first is that explicit and quantitative targets or objectives can be set and achieved in line with quantitative policy guidelines (e.g. Australia is committed to the protection of representative ecosystems and to the protection of rare and endangered species). For example, a set of targets might be to conserve 15% of each ecosystem type (as in the Australian Regional Forest Agreements). The equivalent index-based approaches can only set targets such as to conserve the largest, most biodiverse and/or rarest areas, which tells us nothing about the overall amounts of each asset that will end up in our final set of conservation priority areas. Other objectives in systematic methods can be framed to promote the persistence of biodiversity processes (Pressey *et al.*, 2007) or to minimize landholder opportunity costs. Without explicit objectives and targets, index-based approaches struggle to deal with these kinds of trade-offs.

The second advantage is enhanced complementarity and efficiency. As the whole of a conservation area system is worth more than the sum of the parts, the systematic approach aims to select areas that complement each other and the existing network in terms of the conservation assets. On the other hand, scoring approaches assess each area individually. Accounting for spatially variable information on the cost of specific actions substantially improves efficiency, compared with the approach of designating 'priority areas' and considering actions and their costs *post hoc* (Carwardine *et al.*, 2006).

The third benefit is that systematic conservation planning tools generate multiple alternative sets of areas that meet conservation objectives, providing flexible options and measures of irreplaceability. Irreplaceability can be used as a quantitative measure of priority; areas with higher irreplaceability are likely to require more urgent action because, if they are lost, targets for one or more biological assets are unable to be met. Conversely, higher scores in index-based systems do not necessarily equate to a required urgency of action to protect assets, because the scores were not derived using asset-based targets (Pressey, 2002).

The final advantage relates to adequacy and persistence. Adequacy refers to the existence of sufficient redundancy to provide 'insurance' in the event of extreme events, together with sufficient resources to ensure the persistence of biodiversity processes such as natural patch dynamics and catchment processes. Adequacy is difficult to quantify and implement, in both scoring and systematic approaches. However, systematic methods are being developed to achieve explicit objectives related to adequacy (Pressey *et al.*, 2007). Some of these are being adapted specifically for freshwater systems, for example by considering longitudinal links with upper reaches and lateral connectivity to the floodplain (Simon Linke, pers. comm.).

Examples of systematic approaches in New Zealand and Australia

While scoring approaches have their uses, an important long-term goal should be the development of systematic networks of protected freshwater ecosystems, using a variety of protection techniques including the use of protected areas (Saunders *et al.*, 2002) as well as off-reserve protection (Whitten *et al.*, 2002). Habitat condition has long been used in scoring freshwater systems (Ladson *et al.*, 1999; Norris *et al.*, 2001) and conservation planning software is being modified to incorporate flow, condition and other freshwater-specific considerations. A transition to a nationally adopted systematic approach for establishing conservation priorities in freshwater ecosystems is feasible, logical and efficient.

One example is the New Zealand Waters of National Importance Project (Chadderton *et al.*, 2004). These authors identified New Zealand's nationally important water bodies for biodiversity protection by combining environmental and biogeographic frameworks with information about the distribution of threatened species and communities, and a range of human pressure variables that collectively indicate the naturalness of the system. They argued that if representative and ecologically viable units of the full range of environments and species within each biogeographic unit are protected, then it should be possible to conserve a full range of what remains of New Zealand's freshwater biodiversity.

Chadderton *et al.* (2004) based their approach on three principles. First, least disturbed waters have retained most natural biodiversity and are therefore the highest priorities for protection if further loss is to be minimized. Second, all wetland environment types must be represented among those protected to retain the full range of natural habitats and ecosystems. Finally, remaining threatened native species or community types, where known, also need to be protected, so that viable populations of all indigenous species and subspecies can be maintained. Their approach underlines the importance of criteria related to disturbance, representation and special values, and emphasizes a systematic approach incorporating complementarity.

To do this, sites of 'national importance' in the context of freshwater biodiversity were identified, comprising a minimum area that could protect at least one example of every distinct freshwater ecosystem – allowing that such a list would be enlarged if necessary to include habitats of endangered species. This produced a list which was unacceptably long when assessed against judgements about what was likely to be politically feasible, especially with regard to rivers. Chadderton *et al.* (2004) stated that to capture 100% of distinct river system types would require a major expansion of the identified areas, so compromises were suggested to reduce the number of sites and their total area. The New Zealand approach has much to recommend it, in terms of logic and science. It specifies a scientifically defensible goal, and sets out to achieve it using a comprehensive examination of national ecosystems. Selection rules are transparent, and quantitative indicators are used wherever possible.

Australia is approaching a situation where sufficient data will exist at a national scale to apply this powerful technique. Recently, Norris *et al.* (2007) assessed the condition of 210 000 km of Australian rivers in little more than a year (Australia has just under 3 million km of mapped rivers and streams). Their approach was driven by a hierarchical model of river functioning, which assumed that broad-scale catchment characteristics affect local hydrology, habitat features, water quality and, ultimately, aquatic biota. This model was the basis for selecting important ecologically relevant features that indices should represent. For each reach of each river, a biological index and an environmental index was derived, based on measures quantifying catchment and hydrological condition, and habitat and water quality condition. Data came from existing state and national databases, satellite images, site measurements and process models. The audit classified 14% of reaches as largely unmodified, 67% as moderately modified and 19% as substantially modified by human impacts (Norris *et al.*, 2007).

The team found that when they integrated the three measures of irreplaceability, condition and vulnerability, two major groups of catchments requiring urgent conservation emerged. One of these (7% of reaches) was highly irreplaceable, highly vulnerable but in degraded condition, and so was flagged for restoration. The second group was highly irreplaceable catchments in good condition but vulnerable (2.5% of reaches), and these became priority areas for assigning river reserves (Linke *et al.*, 2007).

Conclusions and prognosis

For Australian riverine and subterranean fresh waters, protection of discrete areas in terrestrial reserves will be unsuccessful if adjacent sub-catchments are being degraded, upstream waters are dammed or extracted, weirs downstream stop the normal migration of fish and groundwater recharge areas are affected by urban development or changed land use. Ideally, aquatic protected areas need to be part of protected landscapes. Given the spatial extent and dynamic nature of most fresh waters, reserves will also need to be large enough to integrate natural patterns of change (Saunders *et al.*, 2002).

According to the Convention on Biological Diversity 1992, the conservation of biodiversity, including aquatic biodiversity, requires the protection of representative examples of all major ecosystem types, coupled with the sympathetic management of ecosystems outside those protected areas. Although all state governments and the Australian Commonwealth government are committed to this principle, only Victoria, Tasmania and the ACT have funded specific programmes to establish representative networks of inland aquatic protected areas. In Victoria and Tasmania, these systems remain incomplete after implementation delays of many years. Although all jurisdictions have national parks and Ramsar sites that protect aquatic ecosystems contained within them, the degree to which such reserves protect *representative* aquatic ecosystems has not yet been systematically assessed in any Australian state. Given the extent of surface and subterranean hydrological connectivity, there is no doubt that many important Australian river and subterranean ecosystems remain with little or no effective protection in spite of these reserves (Stein, 2006).

The evaluation of Australian fresh waters for nature conservation involves techniques for assessing value, importance, condition and threat (Dunn, 2004). Such approaches are under development in Australia, especially the assessment of condition (Brierley & Fryirs, 2000; Norris *et al.*, 2001, 2007). Comprehensive state or national inventories of inland aquatic ecosystems are urgently needed to enable identification and selection of protected areas, and to support recent regional resource planning initiatives (Kingsford & Nevill, 2006). An issue now on the Australian government agenda is the development of a national framework for the establishment of comprehensive, adequate and representative aquatic protected areas (Natural Resource Management Ministerial Council, 2004). The protection of rivers with high conservation value has been under discussion between the states and the Australian government, so far with little sign of concrete progress. Subterranean fresh waters and GDEs are only just starting to draw attention, and will pose their own particular problems in assessments of value, importance, condition and threat (Boulton, 2005). Decade-long delays in implementing official government groundwater policy exacerbate existing ecosystem threats from climate change and over-allocation of both surface and groundwaters for human use (Nevill, 2007a).

Australia has an opportunity to halt the degradation of freshwater ecosystems in the southern half of the continent while conserving relatively undisturbed ecosystems in the tropical north. Success requires implementation of at least five key strategies:

1. Application of adaptive management approaches within regional natural resource management plans (already required by current bilateral natural resource management agreements).
2. Development of comprehensive, accessible and current ecosystem inventories to inform such plans (Nevill & Phillips, 2004).
3. Systematic development of comprehensive, adequate and representative systems of freshwater protected areas, including rivers and subterranean ecosystems (Kingsford & Nevill, 2006).
4. Widespread, appropriate application of the precautionary principle (to which all Australian governments are theoretically committed by policy statements).
5. Recognition of difficulties inherent in the control of the cumulative impacts of incremental development, and a thoughtful search for effective management solutions (Finlayson *et al.*, 2008).

While there is a need to refine understanding of aquatic ecosystem function, Australian freshwater managers and scientists have enough information to chart a way forward. The necessary policies and statutes are also in place – although an Australian version of the Canadian Heritage Rivers System could be an important initiative to coordinate catchment conservation efforts and to mobilize community action (Kingsford, 2007). The issue now is a question of action; future generations will be the judge.

Acknowledgements

We are grateful for the invitation to contribute this chapter and to draw on the experiences of the other authors of this book. We thank Julian Prior, Michael Douglas and Phil Boon for constructive comments on an earlier draft of this paper. Our discussion of systematic conservation approaches draws from a manuscript in preparation by Simon Linke, Bob Pressey and Josie Carwardine, and we thank them for access to this material. We also thank Jim Tait, John Koehn, Naomi Rea, David Hohnberg and various members of the Australian Society for Limnology for useful discussion.

References

ARMCANZ Agriculture and Resource Management Council of Australia and New Zealand, and the Australian and New Zealand Environment and Conservation Council ANZECC (2000). *National Water Quality Management Strategy: Australian and New Zealand Guidelines for Fresh and Marine Water Quality, Volume 1, The Guidelines* (Chapters 1–7). Canberra: Australian Government Publishing Service.

Arthington, A. H. & Pusey, B. J. (2003). Flow restoration and protection in Australian rivers. *River Research and Applications*, **19**, 377–95.

Australian Tropical Rivers Group (2004). *Securing the North: Australia's Tropical Rivers.* Sydney: Worldwide Fund for Nature (WWF Australia).

Bennett, J., Sanders, N., Moulton, D. *et al.* (2002). *Guidelines for Protecting Australian Waterways.* Canberra: Land and Water Australia.

Boulton, A. J. (2005). Chances and challenges in the conservation of groundwater-dependent ecosystems. *Aquatic Conservation: Marine and Freshwater Ecosystems*, **15**, 319–23.

Boulton, A. J. & Brock, M. A. (1999). *Australian Freshwater Ecology: Processes and Management.* Adelaide: Gleneagles.

Boulton, A. J., Humphreys, W. F. & Eberhard, S. M. (2003). Imperilled subsurface waters in Australia: biodiversity, threatening processes and conservation. *Aquatic Ecosystem Health and Management*, **6**, 41–54.

Brierley, G. J. & Fryirs, K. (2000). River styles, a geomorphic approach to catchment characterization: implications for river rehabilitation in the Bega Catchment, New South Wales. *Environmental Management*, **25**, 661–79.

Brierley, G. J., Fryirs, K., Outhet, D. & Massey, C. (2002). Application of the River Styles framework as a basis for river management in New South Wales, Australia. *Applied Geography*, **22**, 91–122.

Carwardine, J., Wilson, K. A., Watts, M. E. *et al.* (2006). *Where Do We Act to Get the Biggest Bang for Our Buck? A Systematic Spatial Prioritisation Approach for Australia.* Report to the Department of the Environment and Heritage, Department of the Environment and Heritage, Canberra, Australia.

Chadderton, W. L., Brown, D. J. & Stephens, R. T. (2004). *Identifying Freshwater Ecosystems of National Importance for Biodiversity – A Discussion Document.* Department of Conservation, Wellington, New Zealand.

Commonwealth of Australia (1992a). *National Strategy for Ecologically Sustainable Development.* Canberra: Australian Government Publishing Service.

Commonwealth of Australia (1992b). *Intergovernmental Agreement on the Environment: An Agreement by the Council of Australian Governments.* Canberra: Department of Environment, Sport and Territories.

Commonwealth of Australia (1996). *National Strategy for the Conservation of Australia's Biological Diversity.* Canberra: Department of the Environment, Sport and Territories(www.erin.gov.au/net/biostrat.html).

Commonwealth of Australia (2003). *National Water Initiative: A Discussion Paper.* Canberra: Department of Prime Minister and Cabinet.

Conference of the Parties (2004). *Report from the 7th Conference of Parties to the Convention on Biological Diversity: Resolution VII/4: Biological Diversity of Inland Water Ecosystems.* United Nations Environment Program website: www.biodiv.org/.

Crabb, P. (1997). *Murray-Darling Basin Resources.* Canberra: Murray-Darling Basin Commission.

Dunn, H. (2000). *Identifying and Protecting Rivers of High Ecological Value.* Occasional Paper 01/00. Canberra: Land and Water Resources Research and Development Corporation.

Dunn, H. (2004). Defining the ecological values of rivers: the views of Australian river scientists and managers. *Aquatic Conservation: Marine and Freshwater Ecosystems*, **14**, 413–33.

Environment Australia (2001). *Directory of Important Wetlands in Australia*, 3rd edn. Canberra: Environment Australia.

Evans, A. G. (2001). *C.Y. O'Connor: His Life and Legacy.* Perth: University of Western Australia.

Evans, R. (2007). *The Impact of Groundwater Use on Australia's Rivers: Technical Report*, Land and Water Australia, Canberra.

Finlayson, B., Nevill, J. & Ladson, A. (2008). *Cumulative Impacts in Water Resource Development.* Paper presented at the Water Down Under Conference, Adelaide, 14–18 April 2008; International Centre of Excellence in Water Resources Management.

Government of Victoria (2002). *Healthy Rivers, Healthy Communities and Regional Growth: Victorian River Health Strategy.* Melbourne: Department of Natural Resources and Environment.

Hatton, T. & Evans, R. (1998). *Dependence of Ecosystems on Groundwater and Its Significance to Australia.* Canberra: Land and Water Australia.

Higgins, J. V., Bryer, M. T., Khoury, M. L. & Fitzhugh, T. W. (2005). A freshwater classification approach for biodiversity conservation planning. *Conservation Biology*, **19**, 432–45.

Kingsford, R. (2007). *Heritage Rivers: New Directions for the Protection of Australia's High Conservation Value Rivers, Wetlands and Estuaries*. Sydney: University of New South Wales School of Biological Sciences.

Kingsford, R. and Nevill, J. (2006). Urgent need for a systematic expansion of freshwater protected areas in Australia: a scientists' consensus statement. *Pacific Conservation Biology*, **12**, 7–14.

Kingsford, R. T., Dunn, H., Love, D. *et al.* (2005). *Protecting Australia's Rivers, Wetlands and Estuaries of High Conservation Value: A Blueprint*. Canberra: Department of the Environment and Heritage.

Kirkpatrick, J. B. (1983). An iterative method for establishing priorities for the selection of nature reserves: an example from Tasmania. *Biological Conservation*, **25**, 127–34.

Koehn, J. (2001). The impacts of weirs on fish. In *The Way Forward on Weirs*, eds. S. Blanch, S. Newton & K. Baird. Sydney: Inland Rivers Network, pp. 59–66.

Ladson, A., Doolan, J., White, L., Metzeling, L. & Robinson, D. (1999). Development and testing of an Index of Stream Condition for waterway management in Australia. *Freshwater Biology*, **41**, 453–68.

Lake, P. S. (1995). Of floods and droughts: river and stream ecosystems of Australia. In *River and Stream Ecosystems*, eds. C. E. Cushing, K. W. Cummins & G. Minshall. Amsterdam: Elsevier, pp. 659–94.

Lake, P. S. (2001). On the maturing of restoration: linking ecological research and restoration. *Ecological Management and Restoration*, **2**, 110–15.

Lake, S. (2005). Perturbation, restoration and seeking ecological sustainability in Australian flowing waters. *Hydrobiologia*, **522**, 109–20.

Lake, S., Bond, N. & Reich, P. (2007). Linking ecological theory with stream restoration. *Freshwater Biology*, **52**, 597–615.

Linke, S., Pressey, R. L., Bailey, R. C. & Norris, R. H. (2007). Management options for river conservation planning: condition and conservation revisited. *Freshwater Biology*, **52**, 918–38.

Margules, C. R. & Pressey, R. L. (2000). Systematic conservation planning. *Nature*, **405**, 243–53.

National Land and Water Resources Audit (2001). *Australian Water Resources Assessment 2000*. Canberra: Commonwealth of Australia. (audit.ea.gov.au/ANRA/water/docs/river_assessment/River_assessment.pdf).

Natural Resource Management Ministerial Council (2004). *Directions for the National Reserve System – A Partnership Approach: Draft for Public Comment*. Canberra: Department for the Environment and Heritage.

Nel, J. L., Roux, D. J., Maree, G. *et al.* (2007). Rivers in peril inside and outside protected areas: a systematic approach to conservation assessment of river ecosystems. *Diversity and Distributions*, **13**, 341–52.

Nevill, J. (2006). *Counting Australia's Protected Rivers*, OnlyOnePlanet Australia (http://www.ids.org.au/~cnevill/FWPA_protectedRivers.htm).

Nevill, J. (2007a). *Groundwater Reform in the Murray-Darling Basin: An Example of Cumulative Impact Policy Failure*, OnlyOnePlanet Australia (www.ids.org.au/~cnevill/FW_MDB_ groundwaterReform.doc).

Nevill, J. (2007b). Policy failure: Australian freshwater protected area networks. *Australian Journal of Environmental Management*, **14**, 35–47.

Nevill, J. & Phillips, N. (2004). *The Australian Freshwater Protected Area Resourcebook*, OnlyOnePlanet Australia www.ids.org.au/~cnevill/FW_ProtectedArea_Sourcebook.doc.

Norris, R. H., Prosser, I., Young, B. *et al.* (2001). *The Assessment of River Condition (ARC). An Audit of the Ecological Condition of Australian Rivers.* Canberra: National Land and Water Resources Audit Office.

Norris, R. H., Linke, S., Prosser, I. *et al.* (2007). Very-broad-scale assessment of human impacts on river condition. *Freshwater Biology*, **52**, 918–38.

Palmer, M. A., Bernhardt, E. S., Allan, J. D. *et al.* (2005). Standards for ecologically successful river restoration. *Journal of Applied Ecology*, **42**, 208–17.

Pressey, R. L. (2002). The first reserve selection algorithm – a retrospective on Jamie Kirkpatrick's 1983 paper. *Progress in Physical Geography*, **26**, 434–41.

Pressey, R. L. & Taffs, K. H. (2001). Scheduling conservation action in production landscapes: priority areas in western New South Wales defined by irreplaceability and vulnerability to vegetation loss. *Biological Conservation*, **100**, 355–76.

Pressey, R. L., Cabeza, M., Watts, M. E., Cowling, R. M. & Wilson, K. A. (2007). Conservation planning in a changing world. *Trends in Ecology and Evolution*, **22**, 583–92.

Rutherfurd, I. D., Jerie, K., & Marsh, N. (2000). *A Rehabilitation Manual for Australian Streams. Volume One: Concepts and Planning. Volume Two: Rehabilitation Tools.* Canberra: Land and Water Australia.

Rutherfurd, I. D., Jerie, K., Walker, M. & Marsh, N. (1999). Don't raise the Titanic: how to set priorities for stream rehabilitation. In *The Challenge of Rehabilitating Australia's Streams*, eds. I. Rutherfurd & R. Bartley. Melbourne: Cooperative Research Centre for Hydrology, pp. 527–32.

Rutherfurd, I. D., Ladson, A. R. & Stewardson, M. J. (2004). Evaluating stream rehabilitation projects: reasons not to and approaches if you have to. *Australian Journal of Water Resources*, **8**, 57–67.

Saunders, D. L., Meeuwig, J. J. & Vincent, A. C. J. (2002). Freshwater protected areas: strategies for conservation. *Conservation Biology*, **16**, 30–41.

Schofield, N. J., Collier, K. J., Quinn, J., Sheldon, F. & Thoms, M. C. (2000). River conservation in Australia and New Zealand. In *Global Perspectives on River Conservation: Science, Policy and Practice*, eds. P. J. Boon, B. R. Davies & G. E. Petts. Chichester: John Wiley, pp. 311–33.

Smith, D. I. (1998). *Water in Australia. Resources and Management.* Melbourne: Oxford University Press.

Smith, P. G. R. & Theberge, J. B. (1987). Evaluating natural areas using multiple criteria: theory and practice. *Environmental Management*, **11**, 447–60.

Sowa, S. P., Gust, A., Morey, M. E. & Diamond, D. D. (2007). A gap analysis and comprehensive conservation strategy for riverine ecosystems of Missouri. *Ecological Monographs*, **77**, 301–34.

Stein, J. (2006). *A Continental Landscape Framework for Systematic Conservation Planning for Australian Rivers and Streams.* PhD thesis, Australian National University, Canberra.

Wentworth Group of Concerned Scientists (2002). *Blueprint for a Living Continent: A Way Forward.* Sydney: Worldwide Fund for Nature (WWF Australia).

Wentworth Group of Concerned Scientists (2003). *Blueprint for a National Water Plan.* Sydney: Worldwide Fund for Nature (WWF Australia).

White, L. J., Rutherfurd, I. D. & Hardie, R. E. (1999). The cost of stream management and rehabilitation in Australia. In *The Challenge of Rehabilitating Australia's Streams*, eds. I. Rutherfurd & R. Bartley. Melbourne: Cooperative Research Centre for Hydrology, pp. 697–703.

Whitten, S., Bennett, J., Moss, W., Handley, M. & Phillips, B. (2002). *Incentive Measures for Conserving Freshwater Ecosystems.* Canberra: Department of the Environment and Heritage Australia.

11 · *Evaluating fresh waters in South Africa*

JAY O'KEEFFE AND CHRISTA THIRION

Introduction

This chapter describes the policies, approach and methods used to assess the conservation value of fresh waters in South Africa. In it we have tried to describe the geographical, historical and socio-economic context which have together resulted in marked differences between South Africa and many other regions of the world – and particularly Europe and the USA.

Climate and topography

South Africa is largely a semi-arid region. In a country which is blessed with an abundance of many natural resources and minerals, water has been a major limiting factor for economic development. Average precipitation in South Africa is 500 mm per year, but high evaporation rates result in runoff of only 8%. The climate in South Africa is also unpredictable and highly variable, so months or years of drought reduce the assurance of the water supply.

Main population centres in South Africa have developed around concentrations of mineral deposits, producing a geographical mismatch between water availability and water need. Gauteng Province (incorporating Johannesburg, Soweto and Pretoria), for example, contains about 17% of the country's 44 million inhabitants, but sits on several catchment boundaries, with all the local rivers draining away from it. The popular management solution to this mismatch has been to redirect water from catchments with a relative abundance but less demand, to those with a shortage. These inter-basin transfers have had major effects on the ecological integrity of both donor and recipient rivers (Davies *et al.*, 1992).

As a result of recent tectonic activity, the central area of the South African land mass is elevated, with the result that most rivers are deeply incised into the landscape. Because of this, and the general aridity, there

Assessing the Conservation Value of Fresh Waters, ed. Philip J. Boon and Catherine M. Pringle. Published by Cambridge University Press.

are very few freshwater lakes and floodplains. Most of the water therefore has to be harvested from river channels in steep valleys – in the last 50 years the supply has been augmented by a vigorous programme of river impoundment. This review focuses on rivers, which are the dominant freshwater ecosystems in South Africa.

History and politics

For historical and strategic reasons, direct and indirect subsidies (such as drought relief and government funding of dams) have encouraged farmers to use nearly 60% of all water abstracted in the country for irrigation. Yet, this farming enterprise contributes only 1.4% of the GDP and employs only 1.4% of the country's workforce (South African DWAF, 2002). Previous water legislation linked water rights to land ownership, which was restricted to the 11% of the population who happened to be white.

When the first democratic government was elected in 1994, a major priority was to produce new legislation, which would provide for equitable distribution of water resources. The result was the South African National Water Act, No. 36, which was introduced in 1998. This revolutionary Act recognizes only one water right – 'the Reserve' consisting of two parts: the basic human needs Reserve – the quantity and quality of water needed for human consumption and hygiene allocated at a minimum of 25 L per person per day; and the ecological Reserve – the water required to sustain the goods and services which functioning ecosystems provide. Entitlements to water through land ownership were abolished – under the new Act, all water uses have now to be applied for and licensed.

People and water supply

By 2002 some 7 million people had been provided with basic water services (of the estimated 17 million people who lacked them in 1994), although the infrastructure may be sub-standard in many cases, and there has been a lack of maintenance. Nonetheless, the national government's goal has been to overcome these problems and supply water to the remaining 10 million people within the next two decades (South African DWAF, 2002).

One unfortunate factor has had the effect of alleviating the growing pressure on water supplies: whereas most developing countries are struggling with a high population growth which constantly outruns efforts to overcome widespread poverty, recent predictions in South Africa are that the population will peak at 44 million in 2008, and then decline. This is largely due to the AIDS epidemic (ABT Associates, 2000).

Philosophy and context

The history of conservation in South Africa can best be encapsulated in two issues: Eurocentric values and what are now designated as 'charismatic megafauna'. The concentration on hunting (initially) and then on preserving the remnants of the large animal populations in fenced reserves dominated conservation thinking until the 1960s. This left few options for the conservation of freshwater ecosystems, which are too longitudinal and heterogeneous to be confined in reserves. One or two visionaries, such as Colonel Stevenson Hamilton, the first warden of the Kruger National Park, realized early in the twentieth century that one of the main vulnerabilities of the reserve system was precisely that they did not contain complete catchments, and were vulnerable to the effects of upstream exploitation. The Sabie River, which crosses the Kruger Park from west to east, is a case in point (see Table 11.1).

Early conservation efforts were very much aimed at species conservation, for the benefit and enjoyment of the (white) elite. In freshwater systems, this translated into the introduction and nurture of alien fish species

Table 11.1 *History of the Sabie River in the twentieth century*

Early 1900s	Gold mining in the upper catchment, mine dumps next to the river, arsenic and mercury pollution downstream.
1913	Well sunk at Sabie Bridge to provide drinking water for Park staff, due to pollution of the river.
1922	'The Sabie River virtually changed to a sterile stream' (Col. Stevenson-Hamilton)
1933	Survey by F. B. Jeary indicated that 'micro-organisms' (probably insects and algae rather than bacteria) were absent from the Sabie.
1940s	Mining ceased. Mining Department removes mine dumps.
1950s	1950s onwards: Sabie River recovers to become biologically one of the most diverse in South Africa because tributaries such as the Marite, Mac Mac and Sand were not affected by mining, and acted as refugia for recolonization. There are still some fish species missing from the middle reaches, where cascades prevent upstream recolonization of less mobile species.

NB: Extensive commercial forestry, water abstraction and sedimentation have probably destroyed much of the refuge value of these tributaries today.

for fishing, particularly brown and rainbow trout, and various bass species. Right up until the 1980s, South African provincial conservation authorities were in the anomalous position in which their main freshwater function was the restocking of trout for fly-fishing.

Following pioneering work on the Pongola River (e.g. by Heeg & Breen, 1982) the South African Department of Water Affairs and Forestry (DWAF) took the lead in a policy of balancing water resource development with protection – essentially an early example of aiming for sustainable development. This was particularly evident in proactive attempts to assess the environmental water requirements (EWRs) of rivers and estuaries (Jezewski & Roberts, 1986). Much of this early work was initiated by far-sighted engineers, and by the willingness of the South African freshwater scientific community to engage in a cooperative dialogue with them. This had the very desirable result that, when the water legislation was rewritten in the 1990s, both DWAF and the scientists were in a position to recommend strong environmental measures, with the tools and knowledge to implement them.

Freshwater conservation in action

This section summarizes some of the research and conservation actions that have shaped the present policies for the conservation of freshwater ecosystems in South Africa. Table 11.2 gives some decade-by-decade highs and lows. Much of the activity listed in Table 11.2 provided the background for the drafting of the far-reaching environmental sections of the 1998 South African Water Act. However, the implementation of these aspects has been slow. Chief among the reasons for this is probably a reluctance to change, both among water managers and landowners. The management of water resources has been driven historically from an engineering perspective, with a considerable bias towards farming. There has been a determination to match supply to demand, rather than to match farming activities to available water supplies. This is no longer a dishonest attempt to advantage one section of the community, but appears to be based on a genuine belief that the country should be self-sufficient in food, no matter what the cost. Similarly, the priority given to the provision of basic water supplies and sanitation for people is very understandable, given the number of people who currently have no access to clean water supplies. However, the environmental requirements are often seen as being in direct competition with water for people, rather than as a means of protecting water resources for people to use in the long term.

Table 11.2 *Brief summary of the history of the conservation of freshwater ecosystems in South Africa*

	Systems approach to river research, with classic investigations such as that of Harrison and Elsworth (1958) on the Berg River, Oliff (1960) on the Thukela River, Allanson (1961) on the Jukskei/Crocodile, and Chutter (1963) on the Vaal.
1950s & 1960s	Linking hydrological, geomorphological and ecological processes to gain a functional understanding of rivers, provided the background knowledge that has underpinned the South African conservation effort in rivers.
1970s	Research was diverted to an understanding of the limnology of reservoirs, many of which were by then exhibiting the symptoms of severe pollution and eutrophication.
1980s	Return to ecological research on natural river systems, and a general increase in the understanding of the importance of systems conservation at the catchment scale (reviewed in O'Keeffe, 1989).
	Nature conservation agencies distanced themselves from their previous role as breeders and stockists of alien fish species, and started to identify particular catchments for priority action.
	DWAF began a programme of assessment of environmental flows for rivers, and a policy of waste load allocations to protect water quality.
	Development of conservation assessment methods for rivers, resulting in the expert system-based River Conservation System (O'Keeffe *et al.*, 1987) and the Habitat Integrity method of Kleynhans (1992).
1990s	Kruger National Park Rivers Research Programme: 10-year programme aimed at gaining an understanding of the ecological functioning of the rivers which rise outside the borders of the Park and flow through it.
	New institutional methods for predicting, planning and managing the water resources of the Park, leading to an overhaul of such practices for the South African National Parks as a whole (O'Keeffe & Rogers, 2003; Rogers & O'Keeffe, 2003).
	Acknowledgement of the importance of change and variability in natural ecosystems, leading to a policy of Strategic Adaptive Management.
	Methods for the setting of ecological objectives (Rogers & O'Keeffe, 2003). e.g. objective hierarchies (see example in Table 11.3) provided a way of linking an overall vision, which may be inspiring but unmeasurable, to indicators (such as sampling benthic invertebrates) which are measurable but obscure.
2000s	Operationalizing the 1998 SA Water Act: Assessments of the Reserve, but little implementation. Beginnings of catchment management agencies.

This view of competition rather than complementarity is demonstrably false, as the following example shows: to provide 100 L per person per day for everyone in South Africa for a year would require only one half of a full supply level of the Vaal Dam (one of the larger reservoirs in the country). In other words, it is not water volume that is the impediment to providing people with basic water; it is the cost and logistics of installing the infrastructure (the pipes and taps). In most of the catchments of South Africa, the water resources are more than adequate to provide for domestic water requirements and to maintain the environmental requirements of river systems, and usually the industrial requirements as well. Generally speaking, it is the allocation of water for inappropriate agricultural uses that makes it look as though South Africa has inadequate water resources.

South Africa has recently become a signatory to the Biodiversity Convention, and has passed the National Environmental Management: Biodiversity Act (2004). This Act provides for the creation of the South African National Biodiversity Institute (SANBI) whose duty it will be to manage the planning, monitoring and conservation of South African biodiversity and to promote biodiversity research in a coordinated and aligned fashion. The South African Institute of Aquatic Biodiversity is responsible for the freshwater and coastal water ecosystems. The Biodiversity Act together with the Protected Areas Act and the amendments and regulations published thus far are an extension of the National Environmental Management: Protected Areas Act 2003 (Act 57 of 2003). South Africa has the legislation, the management infrastructure and the expertise to implement far-reaching conservation policies. Some progress is being made, but, as in many countries worldwide, there is a lack of political will to make environmental issues a real priority. This is understandable, in view of the very real and urgent short-term priorities of meeting people's basic needs, but the environmental costs of failing to protect the resources that provide those basic needs will be felt in the long term.

Protecting the best

Since the signing of the Biodiversity Convention and the passing of the National Biodiversity Act in 2004, a National Spatial Biodiversity Assessment Programme has been initiated to identify special areas which should be designated as nodes of representative biodiversity (Driver *et al.*, 2005). The programme is still at an early stage, in which the criteria for special status are being selected, and the country is being classified into different biodiversity regions. As this programme develops it is likely that some rivers will be singled out for special status. Volume 2 of the programme (Nel *et al.*, 2005) deals with rivers, and plans a systematic conservation planning approach, based on representation, persistence and quantitative target setting. The programme is developing river heterogeneity signatures (Roux *et al.*, 2002) defined according to geomorphic provinces (the level one descriptor) and a hydrological index (the level two descriptor). River integrity has been mapped throughout South Africa, to identify the level of departure from natural state of all rivers:

- 29% of mainstem rivers are intact and suitable for conservation purposes;
- 45% are moderately modified;
- 26% are transformed.

Conservation status has been mapped using river heterogeneity signatures. Of the 120 signatures:

- 44% are critically endangered;
- 27% are endangered;
- 11% are vulnerable;
- 18% are least threatened.

Nel *et al.* (2005) concluded that the mainstem rivers of the Berg, Breede, Gouritz and Vaal are the most severely threatened in South Africa, followed by the Olifants/Doring, Great Fish, Crocodile/Marico and Olifants. All these have over 75% of their length in a critically endangered or endangered state, and will need special attention, including development and implementation of catchment management strategies, in order to prevent further loss of biodiversity.

Methods for assessing the conservation value of rivers

In South Africa, the protection of water resources is related to their use, development, conservation, management and control. According to the National Water Act the Minister has the responsibility to develop a classification system for South Africa's water resources. The aim of the classification system is to provide guidelines and procedures for determining different classes of water resources. The classification system is then used to determine the class and resource quality objectives of significant water resources. The aim is to obtain a balance between the need to protect and sustain water resources, and the need to develop and use them.

Previous methods of assessing values for water resources, such as the 'River Conservation System (RCS)' of O'Keeffe *et al.* (1987) provided interesting insights into the priorities of different experts. The RCS is based on an expert system approach, which attempts to capture the expertise and valuation procedures of professional river conservationists. A group of such experts were asked to identify the major questions that they would ask when ascribing a conservation value to a river, and to weight the various questions. Thirty-nine questions were eventually identified and weighted, in three sections – the river, the catchment and the biota. The system provides an overall conservation status percentage score, and a breakdown according to each section and each question. It identifies which questions contributed most to the overall score, and which should therefore be the priorities to address. Such methods are quite detailed, but proved to be more useful in the development as learning curves for practitioners, than in their application. Users of complex models find it difficult to understand (and therefore accept) how the assessment results were arrived at. More recent methods tend to be simpler, making strong use of professional judgement. A measure of habitat integrity was developed by Kleynhans (1992). This is based on a low-altitude helicopter survey, noting impacts along 5 km sections of the river in relation to specified criteria and impact categories. This provides separate assessments for instream and riparian integrity. Standard weightings are applied to each criterion, and overall weighted scores are calculated. These scores are converted to categories A–F (as described for the classification system in Table 11.3), assigned to each 5 km reach.

Table 11.3 *A classification system for rivers*

A Unmodified, natural: resource base has not been decreased or exploited.

B Largely natural with few modifications: resource base decreased to a small extent. Small changes in biota and habitats, but ecosystem functions essentially unchanged.

C Moderately modified: resource base decreased to a moderate extent. Changes in natural biota and habitats, but ecosystem functions still generally unchanged.

D Largely modified: resource base decreased to a large extent. Large changes in biota, habitats and ecosystem functions.

E Seriously modified: resource base seriously decreased, often exceeding acceptable limits. Loss of biota, habitats and functions is extensive.

F Critically modified: resource base permanently exceeds acceptable limits. Complete modification of biota, habitats and functions. In the worst cases, basic ecosystem functions are irreversibly changed.

Another classification system, used to guide the setting of ecological objectives for the Reserve, is the ecological importance and sensitivity (EIS) system, developed by Kleynhans (1999). This relies on expert knowledge and judgement to assess the overall species diversity (instream and riparian); presence of rare and endangered species; endemic species and communities; species and communities intolerant of change; habitat diversity; general biodiversity; connectivity (e.g. importance as a migration route); presence of conservation areas; and the biotic and abiotic fragility and resilience of a system. This is a vital part of the process for setting the future ecological category (EC) (A–D) or high-level objectives for a river. If the river has a low present ecological state (PES), but a high EIS, then the aim would be to improve the state. If the present state is high, and the EIS moderate or low, then the aim would be to maintain the present state.

EWR methods have been designed and tested for South African circumstances since the late 1980s. With the inclusion of EWRs in the form of the Reserve in the National Water Act, specific developments were initiated to tailor-make existing tools to conform to Reserve requirements as well as to develop new methods where necessary. To comply with the requirements for determining the Reserve, EWR assessments must be undertaken within the context of a Reserve determination process. This process consists of an eight-step procedure (see Figure 11.1).

A variety of methods can be applied within each step. However, all of these methods are used to provide two sets of information:

- Determination of the ECs for the present and the future condition (referred to as ecoclassification).
- Determination of the EWRs (quality and quantity) for a series of sites characterizing the river.

The ecological Reserve for a specific river is dependent on the management class set for that river. The first step in determining the ecological Reserve for a specific river is to determine the PES of that river. Based on this, attainable ecological aims and objectives can be set. This information is used within a scenario-based approach, and a range of ecological aims and states therefore have to be considered. For each of these, a flow scenario must be described (Louw *et al.*, 2004).

The results of the process are provided as ECs ranging from A (near natural) to F (completely modified) for the PES and A to D for the recommended EC (see Table 11.3). These categories will be converted

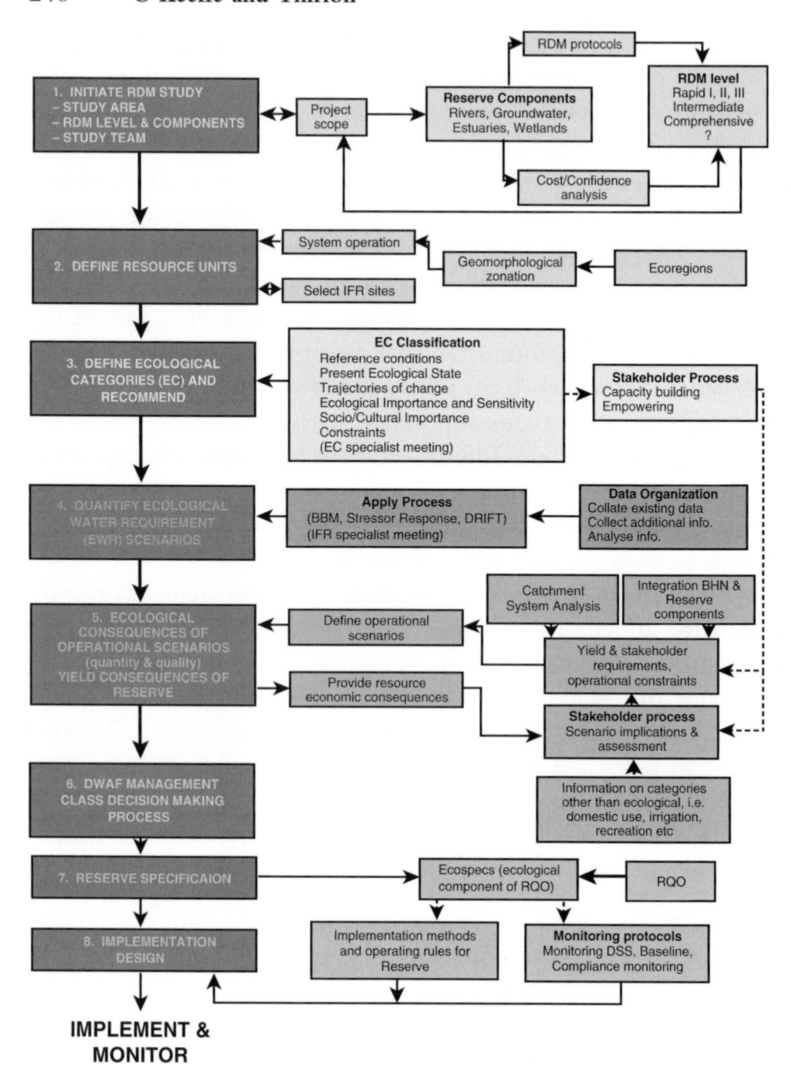

Figure 11.1. The 8-step method for the determination of the ecological Reserve for rivers, from Louw *et al.* (2004).

to more general descriptive terminology when applied to Management Classes that are the output of the classification system procedures still to be devised (Louw *et al.*, 2004).

Several factors are considered when deciding on a recommended EC. These include the PES, importance and sensitivity category, trend (i.e. is the river deteriorating, improving or in a stable state?) and the reason for the PES (i.e. flow alterations, catchment management or introduced alien species). According to the National Water Act all rivers must be managed for a sustainable condition. This means that if the river's PES is a category E or F, management interventions should be made to improve the state to at least a category D. If the river is currently in a sustainable category (A to D), a higher recommended EC will only be considered if the importance category is either High or Very High.

Discussion and conclusions

How to assess the conservation value of freshwater ecosystems

The use of environmental economic techniques has become common in attempts to assign values to freshwater systems. The importance of these valuation techniques has been emphasized by van Wilgen *et al.* (2002):

'In developing countries, where short-term economic growth and social delivery take precedence over conservation, placing a monetary value on ecosystem services is the only way of ensuring intervention'. The question is: How effective are present economic methods for assigning realistic monetary values to natural resources, especially those such as South African rivers, which lack a readily identifiable commercial component, such as a valuable fishery, and in which water – the primary resource – has traditionally been treated as a free good?

In South Africa, and probably elsewhere, environmental economic analysis needs considerable development before it can be effectively used. In particular, there are three areas in which these techniques require improvement. First, present techniques rely on the valuation of components of freshwater ecosystems, rather than a holistic valuation. The most commonly used components are fish, for which there is a market value, reeds and other plant products, the value of the water itself and then contingent valuations of the change in water-borne disease transmission, recreational opportunities, sense of place, etc. These techniques may result in a reasonable valuation where there is a commercial operation, such as in salmon streams in the USA, or white-water rafting on the Zambesi, but they are inadequate for South African rivers, most of which lack these direct commercial assets. These attempts to value systems by reducing them to their components are akin to valuing a car as the sum of its parts: 'Well, the wheels are worth \$100 each; the engine block \$1000; the steering wheel \$120; the wing mirrors \$40, and we're not sure how much the seats are worth; etc.' Any resulting total misses the value of the car as a conveyance. Similarly, there is a need to value freshwater systems in terms of the replacement costs if they were taken out of the landscape. This is difficult, because rivers, for example, usually form the focus of most of the human activity, and without the river, almost no-one would live in the catchment. In this sense the river is more or less priceless. Imagine London without the Thames flowing through it – the reason for London was originally the Thames.

Second, in traditional economic analysis, benefits in the future are discounted, and are therefore worth less the longer-term they are. Economic tools are based on the assumption that a profit now is worth much more than a profit later. Since all environmental benefits are long term, they lose out against anything that makes a quick profit. There needs to be some way of properly valuing the benefits of maintaining natural systems intact over generations, so that our children do not have to bear exorbitant costs for satisfying our immediate priorities.

Third, economic valuations seem always to come back to financial currencies, which are convenient but limited. It would be interesting to see what could be done with the emerging science of measuring human happiness and contentment (Layard, 2005). If such attributes can now be objectively measured, we could start to assess the value of natural resources in terms of the long-term satisfaction that they may give to the people who live by them and depend on them.

Recent advances are beginning to address these issues: Turpie (2004) discusses the concept of 'Total Economic Value of Biodiversity', which includes direct consumptive use value, indirect value (e.g. water supply to downstream users), direct non-consumptive use value (e.g. tourism) and option and existence value. She also discusses the need to use a discount rate that reflects the rights of future generations. Perhaps such developments will result in a more effective role for environmental economics in the future.

The basis of conservation value in South Africa

Because of the historical socio/political/economic conditions in South Africa discussed in the first section, the approach to assessing conservation value in freshwater systems is necessarily very different from that used in Europe and the USA. The main differences are as follows.

Protection for use

The overall priority is more about protection of resources for use than preventing their use for conservation. Although there are glaring examples of unsuitable water uses which should be reduced (e.g. for irrigation in arid areas), in most areas water use will increase in order to give formerly disadvantaged sections of the population equitable access to water for personal and economic benefits. Very few freshwater systems will be exempt from further development, so the emphasis will continue to be on the limits to sustainable development. In a few catchments, primarily those flowing into or through national parks, there will be the chance to make a strong case for limiting use in favour of conservation. However, this should not be seen as the spirit of the 1998 water legislation, which is aimed at overall environmental health, rather than the preservation of remnant areas within fences. There is an increasing view that declaring conservation areas such as national parks is an admission of our failure to implement proper environmental policies throughout the country.

Aridity

Water is generally perceived to be a limiting resource for economic development in South Africa, and therefore is under more pressure for exploitation than in Europe and most of the USA. With modern technology it is quite possible to extract all the water from rivers, using a series of dams and off-stream storage. The major environmental priority for South African rivers is to set limits to water extraction, so that the essential ecological processes and functions can continue. There are, of course, systems in which water quality is the priority issue, particularly in and around heavily urbanized and industrialized areas, and these problems will also have to be addressed.

Poverty

For people who lack basic amenities, the notion of nature conservation is very low down on their list of immediate priorities. Maslow (1967) defined a hierarchy of needs (see Figure 11.2) in which the higher needs are unlikely to become important until the basic needs have been met. In these terms, most of the people of South Africa are still struggling to satisfy the needs of the first three layers, if not elements of the bottom layer (which includes such basics as air, water, food, sleep, sex, etc.) In Maslow's terms, the concept of and necessity for conservation will not become evident until the needs of the lowest four layers of Figure 11.2 have been satisfied. Conservation is firmly in the spiritual and self-actualization layers, which have only become a preoccupation of the citizens of first-world countries. This interpretation presupposes a short-term view,

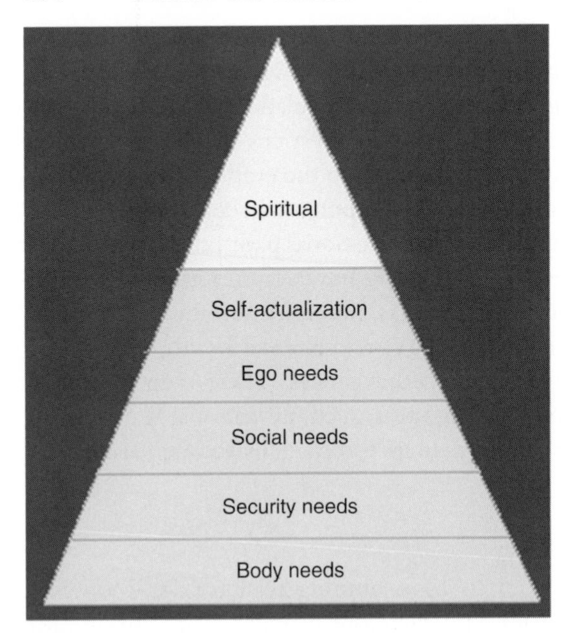

Figure 11.2. Maslow's (1967) hierarchy of needs. Maslow argued that the higher levels of the pyramid are unlikely to become priorities until the more basic needs are satisfied.

which is also the most likely timescale for people who are still struggling to achieve a basic standard of living. It is to the credit of the South African government that they have realized that, in the long-term, the satisfaction of people's basic needs will require the conservation of natural resources, and particularly fresh water, on which the upliftment of the population will depend.

Political history

Related to the last point, the majority of South Africans still view conservation as an elitist preserve of whites, with some justification. The national parks, and the increasing number of private game reserves, are inaccessible to the majority of the population of South Africa, being too expensive and too remote to reach. It must be acknowledged that many conservation organizations are striving to improve accessibility and relevance, by the following means: (1) a system of preferential tariffs for citizens and park neighbours; (2) involving neighbouring communities in using park resources and employment opportunities; (3) encouraging and supporting (free) educational visits from school groups; and (4) appointing full-time staff into the 'social ecology' functions of parks, with a specific mandate to engage neighbouring communities in issues of mutual interest, including river management.

Conservation values

To be successful, conservation cannot be based on values such as rarity, endemism, diversity of species or charismatic megafauna, unless these are also seen as having an economic value. The spectacular growth in the tourism industry, much of it based on wildlife and reserves, is certainly providing an economic impetus for conservation, but the message needs to be more holistic to be successful in the long term. In this regard we return to the theme of protecting natural resources for the use and upliftment of the population in general, rather than the conservation of spectacular remnants of biodiversity for a wealthy elite.

Legislation and policy

In retrospect, although the South African water legislation is much admired around the world, it might have been more effective not to have included specific environmental requirements, such as the Reserve, in the Act itself, but to have described the concept in the Act and confined the details to policy. This is because the inclusion of a statutory requirement for the Reserve as a prerequisite to licensing water use has been a major impediment to implementation. Management agencies are hesitant to agree to statutory environmental flows which they may be unable to implement without major disruption to the existing economic activities and infrastructure in the catchment. For example, in catchments where the water resources have historically been over-allocated, implementation of the Reserve would require reducing the allocations of the users. In rivers with existing dams, the outlet structures may be inadequate to release the environmental flows recommended. These are not insuperable problems, but are daunting for agencies which are already grappling with very serious problems of service provision. More flexible legislation and policies would have allowed a more gradual approach to implementation, which would have been less threatening, and therefore more acceptable, to decision-makers, managers and users.

Appropriate responses to the above issues

If freshwater conservation is to be valued by the majority of South Africans, to the point where they are prepared to forego some immediate economic opportunities to allocate resources to the protection of freshwater ecosystems, then the way that these ecosystems are managed and their conservation promoted must be in terms of their holistic value in the long term. Conservation should be promoted in the sense that it is defined in the water legislation: 'the protection of water resources so that people can use them sustainably in the long term'. Protection measures will have to be carefully motivated in the face of urgent demands for further development and exploitation. The ecological goods and services that natural freshwater resources provide will have to be emphasized. Education and capacity building is essential to empower poor people to be able to look after natural resources, as a prerequisite for breaking out of the poverty trap. Building on the widespread traditional respect and valuation of natural resources (exemplified in religious rituals centred on the spirits of the rivers) is increasingly being emphasized.

As South Africans feel the economic benefits of the tourism boom they will feel less marginalized from national parks and other protected resources. Nature conservation agencies should concentrate major efforts to include surrounding communities in the benefits of eco-tourism. For freshwater ecosystems, education and publicity about the environmental protection aspects of the 1998 Water Act is fundamental to its successful implementation. Proponents of the environmental policies in the Act still have to convince many of the decision-makers and politicians (let alone the general public) that these sections of the legislation should be implemented along with the service provision requirements, and are not in conflict with them. In its first decade of democratic government, South Africa has made brave efforts to promote the conservation value of its freshwater ecosystems – but now it is vital to maintain the momentum into conservation action.

References

ABT Associates (2000). *The State of South Africa's Population 2001/2002.* Report produced for the SA Government Department of Social Development.

Allanson, B. R. (1961). Investigations into the ecology of polluted inland waters in the Transvaal. *Hydrobiologia*, **18**, 1–76.

Chutter, F. M. (1963). Hydrogiological studies on the Vaal River in the Vereeniging area. *Hydrobiologia*, **21**, 1–65.

Davies B. R., Thoms, M. C. & Meador, M. (1992). An assessment of the ecological impacts of inter-basin water transfers and their hidden threats to river basin integrity and conservation. *Aquatic Conservation: Marine and Freshwater Ecosystems*, **2**, 325–49.

Driver, A., Maze, K., Rouget, M. *et al.* (2005). *National Spatial Biodiversity Assessment 2004: Priorities for Biodiversity Conservation in South Africa.* Strelitzia 17. Pretoria: South African National Biodiversity Institute.

Harrison, A. D. & Elsworth, J. F. (1958). Hydrobiological studies on the Great Berg River, western Cape province. Part I. General description, chemical studies and main features of the flora and fauna. *Transactions of the Royal Society of South Africa,* **35**, 125–226.

Heeg, J. & Breen, C. M. (1982). *Man and the Pongolo Floodplain.* South African National Scientific Programmes Report No. 56. Pretoria: CSIR.

Jezewski, W. A. & Roberts, C. P. R. (1986). *Estuarine and Lake Freshwater Requirements.* Technical Report TR 12a. Pretoria: South African Department of Water Affairs and Forestry.

Kleynhans, C. J. (1992). A preliminary assessment of the conservation status of the Lephalala River. In *Instream Flow Requirements of the Lephalala River,* compiler J. M. King. Report to the Department of Water Affairs and Forestry, Pretoria.

Kleynhans, C. J. (1999). Assessment of ecological importance and sensitivity. Appendix R7. Volume 3. In *Water Resources Protection Policy Implementation. Resource Directed Measures for Protection of Water Resources. River Ecosystems.* Version 1.0., ed. H. Mackay. Pretoria: Department of Water Affairs and Forestry.

Layard, R. (2005). *Happiness: Lessons From a New Science.* London: Penguin Books.

Louw, D., Kleynhans, N., Thirion, C., Hughes, D. & O'Keeffe, J. (2004). *Ecoclassification and Habitat Flow Stressor Response Manual.* Report to DWAF, Pretoria.

Maslow, A. H. (1967). *The Farther Reaches of Human Nature.* New York: Viking Press.

Nel, J., Maree, G., Roux, D. *et al.* (2005). *South African National Spatial Biodiversity Assessment 2004. Volume 2: River Component.* Pretoria: CSIR Report Number ENV-S-I-2004–063.

O'Keeffe, J. H. (1989). The conservation of rivers in southern Africa. *Biological Conservation,* **49**, 255–4.

O'Keeffe, J. H. & Rogers, K. H. (2003). River heterogeneity and management outside the Park. In *The Kruger Experience: Ecology and Management of Savannah Heterogeneity,* eds. J. T. du Toit, H. C. Biggs & K. H. Rogers. Washington, DC: Island Press, pp. 447–68.

O'Keeffe, J. H., Danilewitz, D. B. & Bradshaw, J. A. (1987). An expert system approach to the assessment of the conservation status of rivers. *Biological Conservation,* **40**, 69–84.

Oliff, W. D. (1960). Hydrobiological studies on the Tugela River system. *Hydrobiologia,* **14**, 281–392.

Rogers, K. H. & O'Keeffe, J. H. (2003). River heterogeneity: ecosystem structure, function and management. In *The Kruger Experience: Ecology and Management of Savannah Heterogeneity,* eds. J. T. du Toit, H. C. Biggs & K. H. Rogers. Washington, DC: Island Press, pp. 189–218.

Roux, D., de Moor, F., Cambray, J. & Barber-James, H. (2002). Use of landscape-level river signatures in conservation planning: a South African case study. *Conservation Ecology,* **6**, 6 (www.consecol.org/vol6/iss2/art6).

South African Department of Water Affairs and Forestry (DWAF) (2002). *Using Water Wisely. A National Water Resource Strategy for South Africa*. Information Document, August 2002. Pretoria: South African DWAF.

Turpie, J. (2004) The role of resource economics in the control of invasive alien plants in South Africa. *South African Journal of Science*, **100**, 87–93.

van Wilgen, B. W., Marais, C., Magadlela, D., Jezile, N. & Stevens, D. (2002). Win-win-win: South Africa's Working for Water programme. In *Mainstreaming Biodiversity in Development: Case Studies from South Africa*, eds. S. M. Pierce, R. M. Cowling, T. Sandwith & K. MacKinnon. Washington, DC: The World Bank, pp. 5–20.

12 · *Evaluating fresh waters in developing countries*

ROBIN ABELL AND MARK BRYER

Introduction

Freshwater conservation assessments, like their terrestrial and marine cousins, suffer from a geographic imbalance. The world's developed nations have hosted the lion's share of these assessments, yet developing nations are home to a disproportionately large fraction of global freshwater biodiversity (Revenga *et al.*, 1998). This imbalance is entirely understandable and even appropriate. In developed nations, many, if not most, freshwater systems are highly impaired and warrant urgent intervention before species are permanently lost (Dynesius & Nilsson, 1994; Revenga *et al.*, 2000), and those that remain relatively intact are rare gems to be identified and protected with even greater immediacy (Higgins, 2003). In contrast, many of the developing world's freshwater systems have yet to see 'hard-path' infrastructure, aggressive exotics and other near-irreversible impacts, though these stresses are already present in some systems and threaten to affect many more in the not-too-distant future (Wishart *et al.*, 2000; Gleick, 2003; Revenga & Kura, 2003; Nilsson *et al.*, 2005). Opportunities for conservation abound in places like the Amazon, Congo and Mekong, but economic growth pressures do as well. The need for freshwater assessments, then, is equally urgent in these environments, but assessment approaches and outcomes may exhibit important differences.

In the developing world, freshwater conservation planners face two overarching challenges, neither of which is exclusive to developing country contexts but both of which are heightened within them. First, the need to balance biodiversity conservation with economic development interests is a common and strong undercurrent and translates to an elevated focus on freshwater ecosystem goods and services over pure existence values (Wishart *et al.*, 2000). Focusing on the benefits derived from functional freshwater ecosystems, though, is ultimately as much an opportunity as a challenge, because diverse stakeholder groups are by necessity

Assessing the Conservation Value of Fresh Waters, ed. Philip J. Boon and Catherine M. Pringle. Published by Cambridge University Press.

included in the assessment process, making the adoption of results in principle more likely.

Second, data gaps about species, habitats and processes can be so extensive as to nearly engender paralysis (Pringle, 2000). Assessment methodologies created for data-rich situations often transfer imperfectly to these environments, and planners must find creative ways of circumventing data gaps without sacrificing scientific robustness. In some cases, this need has catalyzed technological 'leapfrogging', with a new generation of innovative datasets and tools not requiring *in situ* data being generated and implemented in place of standard tools used within developed countries (Choucri, 1998). Here we present some examples of this phenomenon.

Developed country assessments have played an important role in formulating basic freshwater conservation biology principles and modifying terrestrial approaches to fit the freshwater realm. Developing country assessments are now creating a new dialect within the larger language of freshwater conservation assessments. In essence, the building blocks of the two sets of assessments are the same, but the formulation may be different. The goals of this chapter are to describe these key similarities and differences and to contribute to the dissemination of ideas and tools characteristic of this new generation of assessments.

Philosophy and context

Broad-scale freshwater conservation assessment was born in the developed world. Researchers from Western Europe (principally Great Britain), the United States, Australia and South Africa have all been leaders in the field, generally working independently but generating approaches similar enough to suggest convergent evolution. South Africa, classified as a developing country by some measures, has invested so heavily in systematic and ecological studies, conservation planning and freshwater biodiversity protection that we place it solidly in the developed country column for the purposes of this discussion.

With highly constrained financial resources, the motivation for developing nations to invest in freshwater conservation understandably tilts more towards the provision of ecosystem goods and services than towards biodiversity protection. An increasing recognition of water as a critically limited resource has helped both to draw attention to the need for freshwater assessment and to focus that attention on the role of water in poverty alleviation (United Nations, 2001). Not only is economic well-being a primary consideration for decision makers and stakeholders, but development agencies have explicitly made poverty alleviation a key component of environmental funding (GEF, 1996; World Bank, 2001; USAID, 2003). Biodiversity conservation in the service of poverty alleviation permeates nearly all global discourse; for example, the 2010 Biodiversity Target of the Convention on Biological Diversity aims to achieve 'a significant reduction of the current rate of biodiversity loss at the global, regional and national level as a contribution to poverty alleviation and to the benefit of all life on earth' (Decision VI/26, www.biodiv.org).

While there is as yet incomplete evidence for the contribution of freshwater biodiversity to ecosystem functioning (Gessner *et al.*, 2004), and in turn to services, there is ample support for the direct link between 'healthy' aquatic ecosystems and these benefits (Holmlund & Hammer, 1999; Brismar, 2002). Defining ecosystem health, though, is ultimately subjective (Norris & Thoms, 1999). Ironically, freshwater ecosystems may be considered healthy when they provide ample goods and services, even if those benefits come at the expense of native biodiversity (Allan *et al.*, 2005).

For both practical and philosophical reasons, then, assessing freshwater systems in developing country contexts often involves a greater focus on species of value than on all aquatic taxa, as well as on the biophysical processes that sustain both aquatic species and ecosystem services. Biodiversity is not necessarily lost from the equation, but assessments may be justified and defined in terms of a broader development agenda. The fact that species distribution data are so sparse for most systems makes incorporating biodiversity targets that much harder.

In developed countries, practitioners find the data needed to inform freshwater assessments to be nearly always insufficient, yet data gaps in the developing world make the situation in Western Europe, the United States or other developed regions pale in comparison. As one illustration, nearly 30% of freshwater species in North America were assessed in the 2003 IUCN Red List, whereas in MesoAmerica, South America and the Caribbean islands the numbers were about 7, 6 and 2%, respectively (Revenga *et al.*, 2005). Similarly, in terms of understanding hydrological processes, as of 2002 there were nearly 1000 river gauging stations collecting daily data in North America, yet only 280 such stations in all of South America (data from the Global Runoff Data Centre, http://grdc.bafg.de/). Neither a 30% species assessment level nor a total of 1000 gauging stations is considered sufficient for informing comprehensive assessments, so the substantially lower numbers for developing regions suggest gaps of serious proportions. Even fisheries data, synthesized by the Food and Agriculture Organization, are highly unreliable (R. Welcomme, pers. comm., 2005), and like many datasets their collection on a country-by-country basis renders them difficult to use for basin-wide assessments. Highly valued species may be relatively well studied compared with other taxa, but a recent review of migratory fish of South America suggests that it will be a decade or more before information on migration routes and metapopulation structures will be available for most species (Harvey & Carolsfeld, 2004).

Despite these data gaps, assessments in developing country contexts are under way. They range in extent from continental (e.g. Africa; Thieme *et al.*, 2005), to region-wide (e.g. East Africa; Darwall *et al.*, 2005), to country-wide (e.g. Mexico; Arriaga *et al.*, 2000), to basin-wide (e.g. Congo; Kamdem-Toham *et al.*, 2003), to within-basin (e.g. western Mongolia lakes; Batima *et al.*, 2004) and have included participation by international non-governmental organizations (NGOs) (e.g. WWF, TNC, IUCN, Wetlands International), government commissions and agencies (e.g. Mekong River Commission, Mexico's Comisión Nacional para el Conocimiento y Uso de la Biodiversidad (CONABIO)) and regional (e.g. Nigerian Conservation Foundation) and local organizations. Many of the same groups and individuals who have led assessments in developed country efforts are engaged in these new exercises and in doing so are transferring existing approaches to new settings, where they are then modified to fit local circumstances. The overarching challenge is to assess the conservation value and status of complex, biodiverse systems and to make recommendations that fit with local priorities for sustainable development. Stakeholder participation is widely seen as a critical element if assessment results are ultimately to be adopted (Shumway, 1999).

Protecting the best

Nearly all assessments are motivated to some extent by an interest in protecting the best that a region's freshwater systems have to offer, but many developing country assessments focus primarily on identifying these best examples. 'Best' can be variously defined. When biodiversity conservation motivates an assessment, 'best' might be defined based on the presence of various intrinsic attributes: high species richness or endemism, high habitat diversity, ecological phenomena like large-scale migrations, evolutionary phenomena like relict species or radiations, or rare habitats, species or assemblages (Thieme *et al.*, 2005). When the protection of ecosystem services is paramount, 'best' might refer to those systems whose physical attributes, such as natural hydrographs, remain intact. For degraded systems, 'best' might refer simply to those fresh waters that remain relatively intact, based on direct or indirect measures of biological or physical features.

At the largest scale, a number of international NGOs have in recent years attempted to identify the world's most biologically diverse and representative freshwater systems. Past efforts include WWF's Global 200 (Olson & Dinerstein, 2002), WCMC's global assessment of freshwater biodiversity (Groombridge & Jenkins, 1998) and WRI's Watersheds of the World (Revenga *et al.*, 1998), all of which considered richness and endemism for aquatic groups, with the addition of information on ecological and evolutionary phenomena, in the case of the Global 200, and vulnerability for Watersheds of the World. Continental assessments of developing regions have also been undertaken (Kottelat & Whitten, 1996; Olson *et al.*, 1998; Thieme *et al.*, 2005). Each of these assessments recognized broad data deficiencies and inequalities across regions, with especially prominent gaps in the least developed nations. Additional global and regional evaluations of condition face similar gaps (Revenga *et al.*, 2000; Revenga & Kura, 2003; Nilsson *et al.*, 2005).

Despite data gaps, some freshwater systems rise to the top as obvious examples of exceptional biological value and intactness. Because river discharge is generally correlated with fish species numbers, the world's largest river systems are clear superlatives based on richness (Oberdorff *et al.*, 1995); for example, the Amazon and Congo, with the two highest discharges, also have the two highest numbers of estimated fish species at around 3000 and 700, respectively. In Asia, the Mekong stands out, with the third highest fish richness despite being the 15th largest system globally in terms of discharge (Dudgeon, 2000). All three of these systems also remain remarkably intact across large portions of their basins, are considered global priorities for protection and have been the subject of conservation assessments (Revenga *et al.*, 1998; Baltzer *et al.*, 2001; Olson & Dinerstein, 2002; Kamdem-Toham *et al.*, 2003; WWF, 2005; Nilsson *et al.*, 2005).

Even if their freshwater systems fail to rate as global biodiversity hotspots, all countries have an interest in protecting their own fresh waters. The central importance of functional freshwater systems in providing essential goods and services means that all countries have a responsibility to their citizens to undertake strategic freshwater conservation activities. Furthermore, some freshwater systems may not meet the global bar for biodiversity importance yet still may support regionally or nationally distinct elements (Thieme *et al.*, 2005). Complicating national agendas is the international nature of many freshwater systems; there are at least 263 major transboundary river basins, meaning that most countries will be unable to approach broad-scale freshwater conservation planning and management from a purely national perspective (Yoffe *et al.*, 2004).

In general, the smaller a planning unit, the more feasible it will be to conduct *in situ* surveys for the purpose of identifying the best areas for protection. However, problems of accessibility and funding will still tend to require carefully located sampling with an aim to test hypotheses, rather than geographically comprehensive investigations with the goal of mapping and describing every system. Below, under 'Methods', we discuss typical approaches for identifying biologically significant or relatively intact areas without the benefit of field-derived data.

Responding to environmental threats

Despite the comparatively undisturbed nature of certain freshwater systems within developing countries, in some regions many if not most fresh waters are highly disturbed by deforestation, poor agricultural practices, dams, water withdrawals, overfishing and other impacts. Whereas developed nations altered their freshwater systems in the past and are now focused more on restoration (Bernhardt *et al.*, 2005), alterations are actively under way in many developing nations. In one telling example, Revenga *et al.* (2000), drawing from other sources, reported that 78% of Asia's total reservoir volume was constructed in the 1990s, and that nearly 60% of South America's reservoirs were built since the 1980s. Plans for large dams include 46 in China's Yangtze River Basin, 27 in South America's La Plata and 26 in the Tigris-Euphrates (WWF, 2005). Nilsson *et al.* (2005) found that 47%, 63% and 63% of large river basins in South America, Asia and Africa, respectively, are already either strongly affected or moderately affected by dams and water withdrawals.

Responses to existing threats take a variety of forms and invariably involve community participation. In response to perceived overfishing, local communities in a number of developing countries are experimenting with fishery reserves in floodplain lakes or other systems where fishing can be regulated; prominent examples come from Brazil, Indonesia and Laos (Koeshendrajana & Hoggarth, 1998; de Lima, 1999; Baird, 2000). Conservation assessment can and should contribute to this effort by helping to identify focal habitats, and by establishing robust indicators for monitoring. Traditional ecological knowledge can be a key input to these assessments, as fishers and other local people may have better information than scientists. In one example, fishers in the four lower Mekong riparian countries provided information that, when pieced together, generated a picture of migration routes and spawning periods for 50 fish species (Valbo-Jørgensen & Poulsen, 2000). This kind of information can, in turn, be used to identify potential 'no-go' sites for future dams, or to inform the integration of best management practices into the design and operation of such facilities where they are constructed.

Modification of natural hydrologic regimes takes many forms and is among the most important category of existing and looming threats in developing countries. The assessment of environmental flow requirements – water in rivers and lakes reserved to protect ecological integrity and ecosystem services – is an increasingly prominent response for systems with modified flow regimes, especially where ecosystem goods and services have declined. Environmental flow assessment was born in the developed world, with the majority of environmental flow requirements having been implemented to date in the United States, Australia and South Africa (Tharme, 2003). While flow assessment methodologies used in developed nations tend to be data-intensive and expensive, those used most often in developing countries are designed to work with sparse data, minimal expense and explicit stakeholder participation (Tharme, 2003).

A wide set of responses applied increasingly in developing countries fall under the umbrella of Integrated River Basin Management (IRBM), also known as Integrated Catchment Management, and Integrated Watershed Management. A related approach is Integrated Water Resources Management (IWRM), though, as its name suggests, IWRM is often more concerned with water than with ecosystems. As with most conservation initiatives, IRBM programmes are typically developed in response to a perceived threat, rather than as a proactive measure before ecosystem degradation has occurred. There is also a trend for IRBM programmes in developing countries to focus on objectives tied more to human well-being than to ecosystem integrity; in a review of 35 case studies from around the world, Gilman *et al.* (2004) found that developing country IRBM programmes were significantly more likely than developed country programmes to contain pollution and waste management objectives, as well as agricultural water supply, fisheries management, soil conservation and water storage objectives. Those authors argue that conservation assessments should be an element of all IRBM projects, to enable the investment of scarce resources strategically and in ways that benefit aquatic biodiversity as well as people.

Planning with an eye to forecasted future threats, sometimes referred to as 'futuring', is a promising field that is new to researchers and practitioners working nearly everywhere (Baker *et al.*, 2004; Kepner *et al.*, 2004). Modelling, often central to assessing current conditions in developing countries, is also a key part of forecasting. While inherently uncertain, modelling does have the potential to catalyze conversations about development decisions that might otherwise not take place. In western Mongolia, for example, researchers used a global hydrologic model, coupled with two general circulation models, to project future flow scenarios for rivers under consideration for hydropower dam development. No river gauges exist in the region, but the researchers were nevertheless able to demonstrate that certain river flows would probably decline in the future owing to climate change-related reductions in glacial meltwater, making operation of proposed dams impractical if an environmental reserve were to be maintained (Batima *et al.*, 2004). This kind of creative response simultaneously deals with data gaps and addresses economic considerations.

Evaluating restoration potential

Many freshwater conservation assessments identify restoration priorities, particularly where representation goals cannot be met with intact sites, but restoration potential is often not an explicit prioritization criterion. The Ramsar Convention provides simple guidance for evaluating the restoration potential of wetlands. It recommends that projects be selected in terms of their value both to the environment and to society, and in terms of their feasibility (Ramsar Convention, 2002). Special emphasis is given to evaluating the cost of a restoration project and selecting those with low expense to implementers and landowners. Evaluating the potential of a restoration project in turn requires an analysis of environmental and socioeconomic constraints (Moller, 1999). The Ramsar guidelines emphasize the need not only for socioeconomic assessment but also for stakeholder involvement, and most on-the-ground projects follow this script.

Numerous restoration projects, both large and small, are under way around the world in developing countries. One of particular prominence, now in its early stages, is the restoration of marshlands in Mesopotamia. Following the Ramsar guidelines on restoration, The Iraq Foundation's Eden Again convened a technical advisory panel to undertake the following steps:

1. Review existing information on the Mesopotamian marshlands to evaluate the feasibility of restoration from a scientific perspective and identify major technical challenges.
2. Identify fundamental elements and key ecological and cultural benefits that could be provided through restoration.
3. Conceptualize potential restoration scenarios and identify demonstration projects that would promote recovery of key ecological and cultural benefits.
4. Identify technical considerations and additional data needs for successful restoration efforts.
5. Identify and prioritize processes for increasing the probability that the restoration will successfully achieve its goals (The Iraq Foundation, 2003).

The advisory panel concluded that 'from a scientific perspective, restoration is warranted because it can enable the marshes to provide environmental services, ecological functions, economic goods and socio-cultural values' and that 'restoration efforts are technically feasible and worthwhile'. The group also recommended that

> 'Restoration efforts should reflect the needs and desires of the local population while respecting the importance of the marshes to wider stakeholder interests such as regional and global biodiversity. Local residents and indigenous marsh dwellers will each have different life style and economic requirements in building a new life. In this process of building stakeholder approval and support, mechanisms should be put in place to provide opportunities for local community participation in restoration planning, implementation, monitoring and stewardship' (The Iraq Foundation, 2003).

These recommendations for the Mesopotamian marshes reflect the prevailing sentiment in developing country assessment, both for protection and restoration. Assessment is always goal-driven, and in developing country contexts there is an increasing emphasis on encouraging goal-setting that incorporates the interests of diverse groups. The advisory group in the Mesopotamian effort produced the best scientific assessment possible, but the outcomes were a set of scenarios to enable informed decision-making, rather than a single set of priorities.

Methods for assessing the conservation value of rivers and lakes

Higgins and Duigan (Chapter 4) review a variety of approaches to assess the conservation value of fresh waters, focusing on different taxonomic and geographic scales. Additional detail on methods can also be found in Groves *et al.* (2002), Abell *et al.* (2002), Higgins (2003) and Higgins *et al.* (2005). A summary of this literature yields a generalized methodological framework for assessing the conservation value of fresh waters:

1. Determine the region of analysis.
2. Define conservation targets in the region.
3. Assess the integrity or viability of conservation targets.
4. Set goals for conservation targets.
5. Select areas of high conservation value.

This framework is applicable globally, both in developed and developing regions. Applying this framework in developing countries, however, results in both challenges and opportunities that are often different from those found in assessing conservation value in the developed world. For example, developing countries generally have fewer empirical data about the biodiversity and condition of rivers and streams, causing a heavy reliance on expert knowledge and opinion. Yet some data do exist in nearly every country, and advances in spatial data acquisition, analysis and modelling provide interim surrogates to assess conservation value until more field data can be gathered. With creativity and the recognition that results may be less accurate and/or precise than assessments in developed areas, constructive assessments in developing countries are possible and much can be accomplished using a step-wise, adaptive approach.

Each step of this generalized framework is described briefly below, with examples provided to highlight special considerations and adaptations in developing country contexts.

1. Determine the region of analysis

The first step in assessing conservation value is to determine the region of analysis, typically based on ecological or biodiversity features of interest. There is widespread agreement that catchments, also known as watersheds or drainage basins, are logical management units for freshwater biodiversity and water resource conservation (Wishart & Davies, 2003). However, catchments are nested units defined at a range of scales, so additional information may be needed to identify the appropriate size and to decide if aggregation, or perhaps splitting of catchments, is warranted. For biodiversity conservation assessments, this additional information is typically biogeographic.

The lack of biogeographic data for freshwater species in developing regions has impeded development of assessment units specific to freshwater biodiversity conservation. WWF and TNC have developed a global set of 'freshwater ecoregions', based primarily on the distribution patterns of freshwater fish species; these ecoregions are designed to serve as coarse-scale assessment units that can then be subdivided to capture finer patterns (Abell *et al.*, 2008). Ecoregional boundaries in many poorly sampled developing countries are a best estimate given current knowledge about species distribution patterns, and ecoregion lines may be modified over time as biogeographic knowledge improves.

The Nature Conservancy has developed an approach for subdividing freshwater ecoregions into groups of basins, termed Ecological Drainage Units (EDUs), that share a common zoogeographic history as well as physiographic and climatic characteristics; these basins together represent an area of distinct ecosystem patterns and biotic characteristics (Groves *et al.*, 2002; Higgins *et al.*, 2005). EDUs or other stratification units are used to ensure that conservation targets across environmental gradients and within biogeographic regions are represented in a portfolio of conservation areas (Groves, 2003). Assessments that employ stratification units may be particularly important in tropical developing countries where species turnover may be high; until data are available to test this hypothesis, a conservative approach argues for splitting into greater rather than fewer units, to ensure ample representation within networks of managed areas.

Both biotic and abiotic data inform the development of freshwater ecoregions and EDUs (Abell *et al.*, 2002; Higgins *et al.*, 2005). A fundamental abiotic input is a digital catchment map, derived at a scale appropriate to its use. Until recently, such maps were available only at fixed scales for much of the world owing to limitations in elevation data (see Hydro1K; http://edcdaac.usgs.gov/gtopo30/hydro/). However, newly available 90-m resolution Shuttle Radar Topography Mission (SRTM; www2.jpl.nasa.gov/srtm/) elevation data allow for the generation of catchments at virtually any scale, and WWF is producing a global data set (HydroSHEDS; http://hydrosheds.cr.usgs.gov/). Data needs in developing countries motivated WWF to undertake this project, with a first application in the Madre de Dios River basin in the south-west Amazon, Peru and Bolivia, where derived streams and

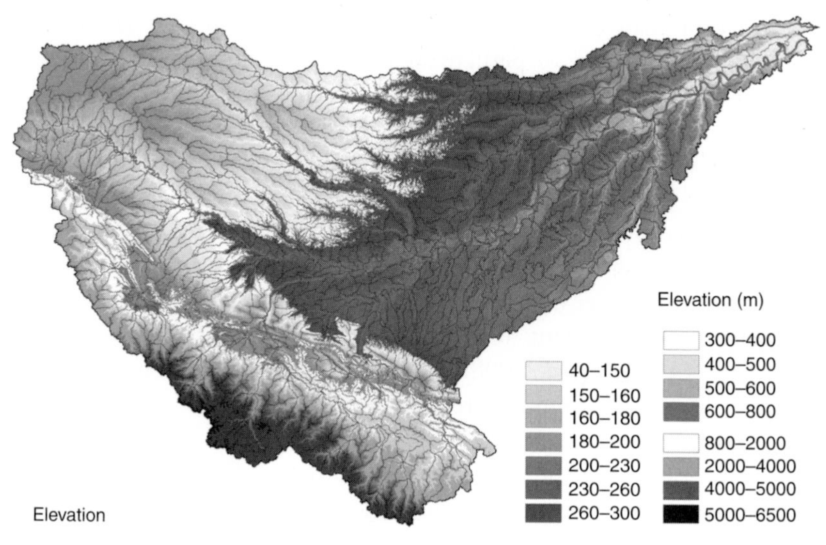

Figure 12.1. An example of catchments for the Madre de Dios and Orthon river basins in Peru and Bolivia. Data from HydroSHEDS (http://hydrosheds.cr.usgs.gov/). (See colour plate)

catchments were required (Figure 12.1). This is a strong example of developing country assessments driving the generation of an advanced tool that will most likely be adopted in developed countries as well. Digital hydrographic data provide the geographic basis for classifying freshwater habitat types, and are also invaluable when assessing system integrity (see step 3 below). Other abiotic inputs such as soils maps, as well as most biotic data, cannot be as reliably modelled as elevation-derived catchments. Planning units can nonetheless be defined, albeit with less certainty than in data-rich situations. In one example, a combination of expert assessment and data analysis was used to develop EDUs in South America's Upper Paraguay (Figure 12.2).

Ecoregions and EDUs are not the only freshwater planning frameworks, and other frameworks can look quite different, despite some shared inputs. For example, Wasson *et al.* (2002) and Navarro and Maldonado (2002) have developed 'hydroecoregions' for Bolivia based on a hierarchy of physiography, geology, bioclimatic patterns and geomorphology. Importantly, these hydroecoregions do not correspond to catchments, and as such they may represent aquatic habitat differences at a greater level of detail than do catchment-based frameworks; all frameworks present trade-offs, and the hydroecoregions may sacrifice ease of application to management for greater precision of habitat mapping.

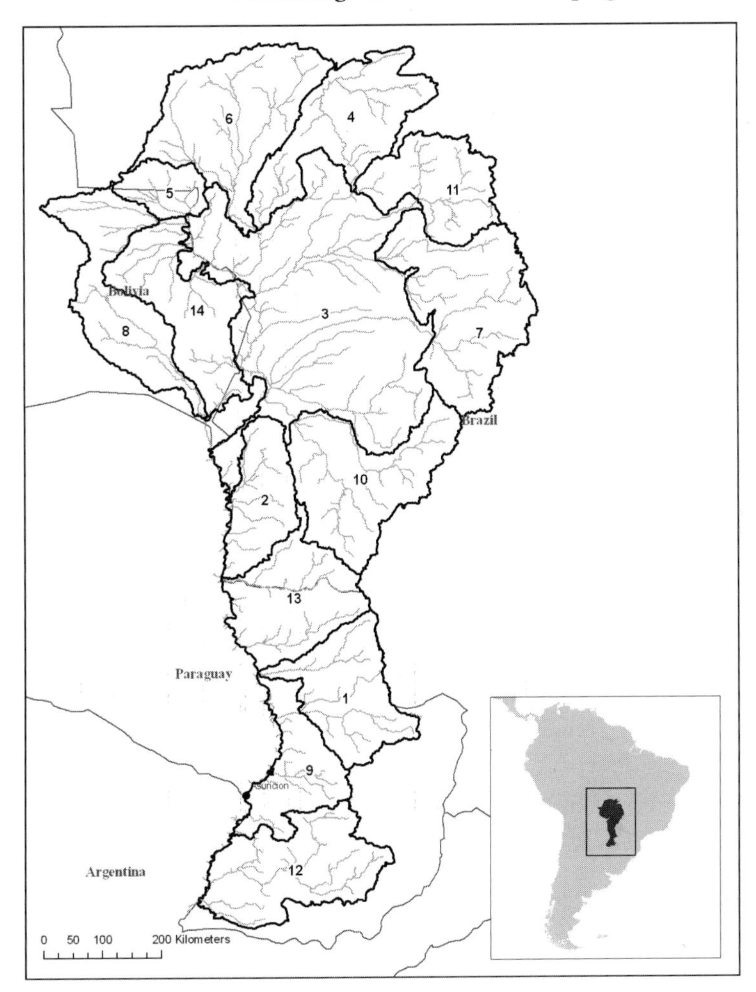

Figure 12.2. Ecological drainage units (EDUs) of the Paraguay River freshwater ecoregion in Brazil, Paraguay and Bolivia, South America. There are 14 EDUs shown for this ecoregion. Adapted and updated from Figure 10.4, chapter 10, in Groves (2003). Copyright © 2003 Island Press.

Finally, many existing efforts to assess conservation value in both developing (PROBIO, 1999) and developed (Master *et al.*, 1998) contexts use political boundaries in determining the region of analysis. Social and political relevance must be considered within any conservation assessment, but ecological boundaries (both inside and outside of a country) are critical at the same time. Transboundary assessments may even provide a catalyst for heightened cross-border cooperation (van der Linde *et al.*, 2001).

2. Define conservation targets in the region

Higgins and Duigan (Chapter 4) describe a hierarchy of biological taxonomy, from genes to species to communities to ecosystems, that can be used to define conservation targets within a region of interest. Regardless of how value is assigned (i.e. which conservation targets are selected), developing countries generally have fewer data and data of lesser quality compared with developed nations (Revenga *et al.*, 2005). Yet, assessments proceed because some data exist everywhere, and surrogates for field data are increasingly available.

At the species level, fish and water birds are generally the best studied species-level taxa in fresh waters throughout the world, with amphibians, freshwater molluscs, crustaceans, insects and plants much more poorly known (Revenga & Kura, 2003). Taxonomies of freshwater fish have been published across most of the major developing regions of the world (Daget *et al.*, 1984; Kottelat, 1989; Reis *et al.*, 2003), but such references quickly need updating as new species are described and taxonomies are revised. During the year 2004, for example, new species of freshwater fish have been described in South America at a rate of approximately four per month (Paulo Petry, pers. comm., 2005). In addition to taxonomic survey gaps, there is also little or no information on the ecology of nearly all aquatic species (Dudgeon, 2000), or on their conservation status (IUCN, 2003). Therefore, even assessments of conservation value that focus only on the subset of species that are most threatened (i.e. coarse-filter/fine-filter approach described in Groves *et al.*, 2002) can be challenging.

Assessments in developing countries that include freshwater species as a measure of conservation value must therefore accommodate these data limitations by supplementing available data with the use of analytical tools. Assessments conducted in the Upper Paraguay Basin in Brazil, for example, measured conservation value using known migratory corridors and spawning habitat for commercially important or well-known taxa (PROBIO, 1999; de Jesus, 2003). An assessment for the lower Mekong basin took a similar approach (Baltzer *et al.*, 2001). These efforts, and those in other data-limited contexts, rely to a large extent on unpublished data and expert knowledge as a primary source of information to identify what is of value and how to assess it. The best expert assessments are cognizant of bias, standardize the way information is collected, and estimate levels of confidence when relying primarily on expert knowledge.

In another example of making the best of existing species information, Terneus *et al.* (2004) identified 104 species of fish as species-level

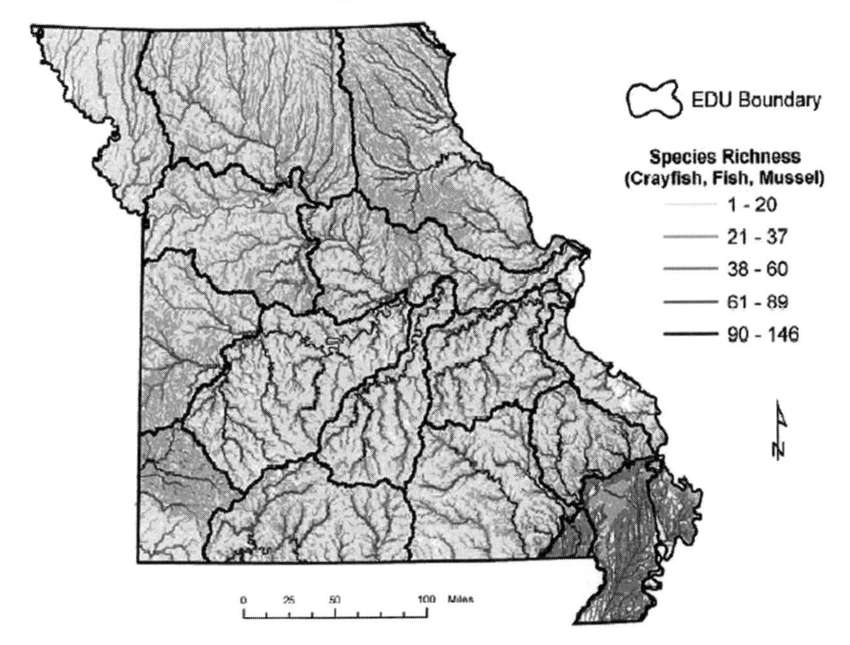

Figure 12.3. Map of freshwater species richness for Missouri, USA. The map is based upon predicted distribution models for 315 fish, mussel and crayfish species using field sampling data from only 0.03% of the streams. Users can also individually select stream segments within a GIS to obtain a list of the species predicted to occur within each segment of interest. From Sowa (2004).

conservation targets in an analysis of Pacific drainages in southern Ecuador and northern Peru. Nearly 700 known occurrences for 90 of these species were obtained from museums, universities and local experts, and each species was classified as to its general habitat preference. Although incomplete, these data provide a way forward for identifying fresh waters of critical value. In some circumstances, enhancement of existing species-level data is also possible through field collection (Chernoff *et al.*, 2003). However, this can be time-consuming and expensive and is generally limited to small-scale efforts.

In areas where sampling has been limited, habitat models can also offer great promise to understand better the distributions of the target species by correlating general habitat characteristics at known locations with the occurrence of those same characteristics across much larger areas (Sowa, 2004; Figure 12.3). Silva and Guevara (2003) developed a similar approach for terrestrial species across the northern Andean range of South America.

Where species data are lacking or require supplementation, many assessments incorporate targets built around the representation of freshwater ecosystem types. Representation of all natural ecosystems is supported in the literature as a proactive conservation approach (Hunter, 1991; Moyle & Yoshiyama, 1994; Angermeier & Schlosser, 1995; Pressey *et al.*, 2000). In cases where species data are especially lacking, freshwater ecosystem diversity and integrity may be the only determinants in assessing conservation value (Roux *et al.*, 2002).

Despite its critical importance, there is no taxonomy of freshwater systems available for most areas in developed and developing countries alike. Higgins *et al.* (2005) describe a method for creating such a taxonomy using available abiotic data and Geographic Information System (GIS) technology. This method is applicable worldwide, but data availability plays an important role in the scale and level of confidence at which a classification of ecosystems can be developed. To some extent, remotely sensed information can serve as a proxy for empirical data and field-based knowledge.

Patterns of community and/or ecological diversity in tropical freshwater systems may differ from those in temperate systems, requiring that models incorporate different assumptions. Classic paradigms developed from experiences in temperate streams, such as the River Continuum Concept (Vannote *et al.*, 1980), may not apply in many tropical scenarios (see Dudgeon, 2000). In addition, certain aquatic habitats are particularly critical to tropical aquatic systems (owing to different hydrologic and climatic conditions) and maintenance of freshwater biodiversity, such as floodplain forests and estuarine areas characterized by mangroves (Pringle, 2000). Modelling efforts must strive to represent these unique habitats despite the lack of precedent in temperate environments.

Recent efforts in Brazil (de Jesus, 2003) and Ecuador and Peru (Terneus *et al.*, 2004) illustrate a range of possibilities available to classify freshwater ecosystems in developing country contexts. In the Upper Paraguay Basin, de Jesus (2003) and Higgins *et al.* (2005) described an approach for ecosystem identification that relied on manual classification and delineation predominantly based on expert knowledge and limited abiotic data (Figure 12.4). In the Pacific drainages of southern Ecuador and northern Peru, Terneus *et al.* (2004) developed an automated classification in GIS of 169 ecosystem types using available digital data for drainage area, altitude, slope, connectivity and soil type, with some field verification. Processed SRTM elevation data, described above, will provide the geographic basis for classification of different freshwater types, particularly when linked to hydrologic models to generate discharge information. Because the data are modelled, however, they will require field verification, often made difficult by the relative inaccessibility of many areas in developing countries.

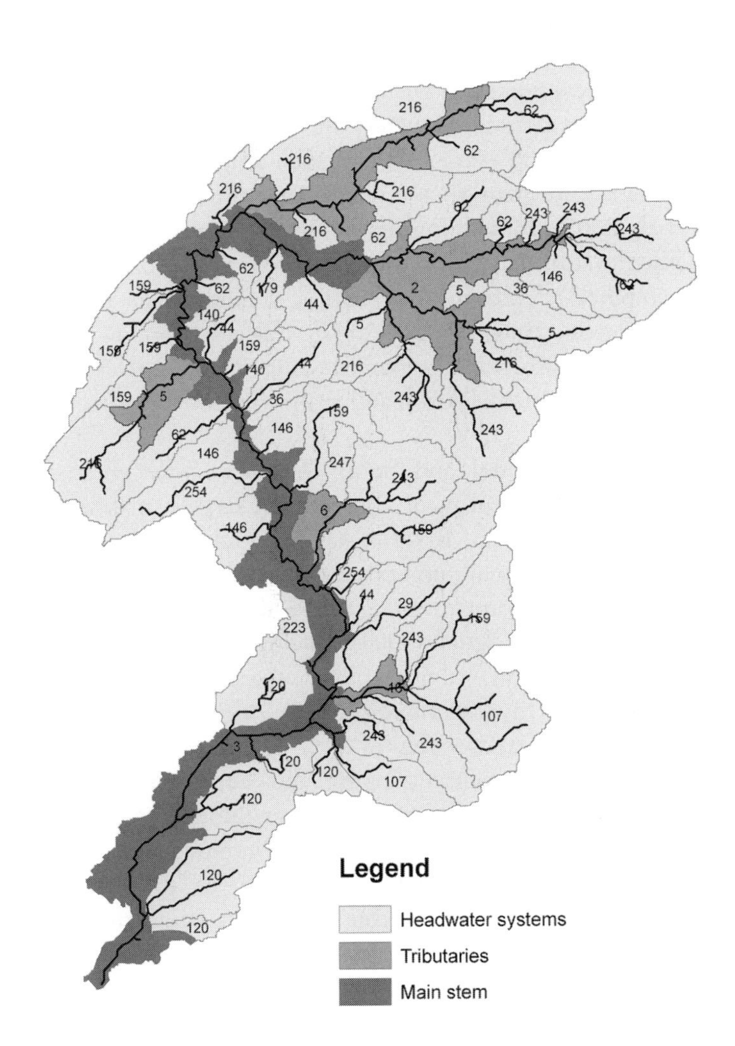

Figure 12.4. Freshwater systems from a portion of the Alto Cuiaba basin in the Paraguay River freshwater ecoregion. Examples of 19 system types are presented as headwater sub-basins, tributaries, and mainstems. Multiple examples are represented as repeated numbers. Adapted and updated from Higgins *et al.* (2005).

3. Assess the integrity or viability of conservation targets

There is a growing literature on assessing the integrity of freshwater conservation targets, reviewed in detail in Chapter 4 and by Higgins (2003) and Abell *et al.* (in press), among others. As with classifications, both direct, field-measured data and indirect data can serve as inputs, with indirect data (and models that may produce them) more likely to be used in developing country assessments. Not only are measured data lacking, but few methods to evaluate biological integrity using field data have been tested adequately in tropical fresh waters. As a result, many of the world's largest rivers, such as the Amazon, Congo and Nile, have no mechanism by which to evaluate their integrity (Revenga *et al.*, 2005). In addition, many developing countries do not have established water quality monitoring programmes from which to understand empirical trends in the physical conditions of freshwater ecosystems (Dudgeon, 2000; Pringle, 2000). Current efforts to develop biological assessment protocols in the Mekong River, however, may provide a template for data collection and analysis in developing countries (Revenga *et al.*, 2005).

Recent efforts in Central and South America show a range of approaches to assessing integrity of conservation targets in developing countries. In Central America, Bryer *et al.* (2001) relied primarily upon land-cover maps and expert opinion to rank qualitatively the integrity of freshwater ecosystems. In comparison, in the Pacific drainages of southern Ecuador and Peru, Terneus *et al.* (2004) developed detailed models for both species and ecosystem viability using automated analysis in GIS. Viability was modelled by combining factors (such as water quality, percentage of natural land-cover in the catchment, presence of introduced species and presence of dams and water diversions) into indices that ranked each example of each target. Note that data for all factors were not available across the entire area of interest, yet a simplified index could still be created. Expert knowledge, be it academic or indigenous, will always play a vital role in understanding the current condition of freshwater systems (Abell *et al.*, in press).

4. Set goals for conservation targets

Goal-setting is intended to define what is required ecologically to achieve conservation success in a region of interest – to answer the question, 'how much is enough?' Goal-setting approaches are similar around the world and typically consist of selecting the number, size, condition (integrity) and spatial distribution (including connectivity) needed for each target in that region to persist into the future (Groves, 2003). Most exercises in goal-setting for aquatic targets come from the developed world, but the science is nascent everywhere, data are lacking, and in all cases the precautionary principle prevails, with goals often set towards the high end (Noss *et al.*, 2002; Smith *et al.*, 2002; Weitzell *et al.*, 2003; Nel *et al.*, 2004). Goals set during assessments need to be viewed as starting hypotheses to be tested into the future.

5. Select areas of high conservation value

Specific criteria that can be used to create a collective portfolio of areas of high conservation value for a region do not differ between developed and developing countries and can include (after Groves, 2003):

- completeness (the areas selected capture the diversity and numbers of targets of conservation value identified in the goals);
- efficiency (the areas selected support multiple targets of conservation value);
- integration (the areas selected consider opportunities beyond freshwater conservation alone);
- functionality (ecological processes in their natural ranges of variability are present or restorable in the areas selected).

Selecting areas of high conservation value clearly requires a synthesis of many layers of information and is best undertaken by a team of local and regional experts interacting at all steps during the process. Either computer-assisted decision support systems (Possingham *et al.*, 1999) or manual selection (Terneus *et al.*, 2004) can be used, although incorporating connectivity as a criterion remains a challenge, particularly for automated processes. Whether or not decisions are made with the help of a computer, stakeholders (e.g. fishermen, farmers) need to play a central role in assessments since they are likely to be affected by, or may even be responsible for implementing, the results. This is even more critical in those countries or regions where governments lack enforcement capability for regulatory decisions.

Conservation planners the world over know the power of maps. Of critical importance to freshwater conservation planning is how priorities are visualized; representing a river reach as a priority by itself may be unsatisfactory in that the need to manage surrounding lands is poorly communicated, yet identifying an entire catchment as a conservation priority can present a frightening image to stakeholders. Having stakeholder buy-in to both the process of setting priorities and the end product is perhaps the best way to ensure that assessments are interpreted correctly, and that they are used optimally to achieve conservation results.

Conclusions

Freshwater conservation assessment and planning in developing countries are often distinguished both by greater challenges and by greater opportunities than parallel efforts in developed country contexts. Biodiversity values are often high, but data describing them are few, and the pressures of impending development may necessitate making conservation recommendations without the benefit of even basic biological information. In many developing countries, a window of opportunity exists now for proactive freshwater conservation interventions, and that window will surely narrow in the future.

Conservation planners working in developed countries are investing in approaches with relatively heavy data requirements. Conservation assessment in developed country contexts will continue to be important, both to conserve freshwater biodiversity in those places and from the perspective of formulating cutting-edge ideas. However, planners should be mindful that the next frontier in freshwater conservation assessment will, by sheer nature of its size, occur in the developing world and that data will be in short supply for some time. Experiments with modifying planning approaches and tools for use in developing country contexts may in fact lead to new sets of unanticipated innovations.

All conservation planning, be it in the terrestrial, freshwater or marine realms, must not sacrifice the perfect for the good. This concept is particularly apt for fresh waters in developing countries, especially considering that biodiversity may be one of many values represented around the table. The relationship of biodiversity or ecosystem conservation to development goals must be addressed frankly, with stakeholder participation critical to achieving buy-in and consensus. Fortunately, there is growing evidence in support of the economic benefits derived from freshwater systems functioning within natural ranges of variation, so ultimately it may be possible to prove that protection will often be in harmony with poverty alleviation goals. Yet we cannot wait for all the evidence to come in, because before long the opportunity for conserving fresh water for the benefit of all life on earth will be lost.

References

Abell, R., Thieme, M. & Lehner, B. (in press). Assessing threats to freshwater biodiversity from humans and human-shaped landscapes. In *Human Population: The Demography and Geography of* Homo sapiens *and their Implications for Biological Diversity*, eds. R. Cincotta & D. Mageean. Heidelberg, Germany: Springer-Verlag.

Abell, R., Thieme, M., Revenga, C. *et al.* (2008). Freshwater ecoregions of the world: a new map of biogeographic units for freshwater biodiversity conservation. *BioScience* **58**: 403–414.

Abell, R., Thieme, M., Dinerstein, E. & Olson, D. (2002). *A Sourcebook for Conducting Biological Assessments and Biodiversity Visions for Ecoregion Conservation. Volume II: Freshwater Ecoregions.* Washington, DC: World Wildlife Fund.

Allan, J. D., Abell, R. A., Hogan, Z. *et al.* (2005). Overfishing of inland waters. *BioScience*, **55**, 1041–1051.

Angermeier, P. L. & Schlosser, I. J. (1995). Conserving aquatic biodiversity: beyond species and populations. *American Fisheries Society Symposium*, **17**, 402–14.

Arriaga, L., Aguilar, V. & Alcocer, J. (2000). *Aguas Continentales y Diversidad Biologica de Mexico.* Mexico D. F., Mexico: Comision Nacional para el Conocimiento y Uso de la Biodiversidad.

Baird, I. G. (2000). *Towards Sustainable Co-Management of Mekong River Inland Aquatic Resources, Including Fisheries, in Southern Lao PDR.* Evaluating Eden Series Discussion Paper No. 15. International Institute of Environment and Development.

Baker, J. P., Hulse, D. W., Gregory, S. V. *et al.* (2004). Alternative futures for the Willamette River Basin, Oregon. *Ecological Applications*, **14**, 313–24.

Baltzer, M. C., Dao, N. T. & Shore, R. G. (eds.) (2001). *Towards a Vision for Biodiversity Conservation in the Forests of the Lower Mekong Ecoregion Complex.* Hanoi, Vietnam and Washington, DC: WWF Indochina/WWF US.

Batima, P., Batnasan, N. & Lehner, B. (2004). *Freshwater Systems of the Great Lakes Basin, Mongolia.* Ulaanbaatar, Mongolia: WWF Mongolia Programme Office.

Bernhardt, E. S., Palmer, M. A., Allan, J. D. *et al.* (2005). Ecology – synthesizing US river restoration efforts. *Science*, **308**, 636–7.

Brismar, A. (2002). River systems as providers of goods and services: a basis for comparing desired and undesired effects of large dam projects. *Environmental Management*, **29**, 598–609.

Bryer, M., Paaby-Hansen, P., Calderon, R. & Perot, J. (2001). *Development and Application of a Preliminary Freshwater Ecosystem Classification in Central America for Use in Regional Conservation Planning.* International Scientific Forum on Integrated Management of Water Resources, First Water Fair of Central America and the Caribbean. Panama City, Panama.

Chernoff, B., Machado-Allison, A., Riseng, K. & Montambault, J. R. (eds.) (2003). *A Biological Assessment of the Aquatic Ecosystems of the Caura River Basin, Bolivar State, Venezuela.* Washington, DC: Conservation International.

Choucri, N. (1998). Knowledge networking for technology leapfrogging. *The Cooperation South Journal*, **2**, 40–52.

Daget, J., Gosse, J. P. & Thys van den Audenaerde, D. F. E. (eds.) (1984). *Check-list of the Freshwater Fishes of Africa.* Belgium: ORSTOM-MRAC.

Darwall, W., Smith, K. & Lowe, T. (2005). *The Status and Distribution of Freshwater Biodiversity in Eastern Africa.* Cambridge, UK: IUCN Species Programme.

de Jesus, F. (Coord.) (2003). *Classificacao dos Ecosistemas Aquaticos do Pantanal e de Bacia do Paraguai.* Brasilia, Brazil: The Nature Conservancy.

de Lima, D. M. (1999). Equity, sustainable development and biodiversity preservation: some questions about ecological partnerships in the Brazilian Amazon. In *Várzea: Diversity, Development and Conservation of Amazonia's Whitewater Floodplains*, eds.

C. Padoch, J. M. Ayres, M. Pinedo-Vasquez & A. Henderson. New York: The New York Botanical Garden Press, pp. 247–63.

Dudgeon, D. (2000). The ecology of tropical Asian rivers and streams in relation to biodiversity conservation. *Annual Review of Ecology and Systematics*, **31**, 239–63.

Dynesius, M. & Nilsson, C. (1994). Fragmentation and flow regulation of river systems in the northern 3rd of the world. *Science*, **266**, 753–62.

GEF (Global Environment Facility) (1996). *Operational Strategy of the GEF.* Washington, DC: The GEF.

Gessner, M. O., Inchausti, P., Persson, L., Raffaelli, D. G. & Giller, P. S. (2004). Biodiversity effects on ecosystem functioning: insights from aquatic systems. *Oikos*, **104**, 419–22.

Gilman, R. T., Abell, R. A. & Williams, C. E. (2004). How can conservation biology inform the practice of Integrated River Basin Management? *Journal of River Basin Management*, **2**, 135–48.

Gleick, P. H. (2003). Global freshwater resources: soft-path solutions for the 21st century. *Science*, **302**, 1524–8.

Groombridge, B. & Jenkins, M. (1998). *Freshwater Biodiversity: A Preliminary Global Assessment.* Cambridge, UK: World Conservation Monitoring Centre.

Groves, C. R. (2003). *Drafting a Conservation Blueprint: A Practitioner's Guide to Planning for Biodiversity.* Washington, DC: The Nature Conservancy and Island Press.

Groves, C. R., Jensen, D. B., Valutis, L. L. *et al.* (2002). Planning for biodiversity conservation: putting conservation science into practice. *BioScience*, **52**, 499–512.

Harvey, B. & Carolsfeld, J. (2004). Introduction: Fishes of the floods. In *Migratory Fishes of South America: Biology, Social Importance and Conservation Status*, eds. J. Carolsfeld, B. Harvey, A. Baer & C. Ross. World Fisheries Trust/World Bank/IDRC, pp. 1–18.

Higgins, J. V. (2003). Maintaining the ebbs and flows of the landscape: conservation planning for freshwater ecosystems. In *Drafting a Conservation Blueprint: A Practitioner's Guide to Planning for Biodiversity*, ed. C. Groves. Washington, DC: The Nature Conservancy and Island Press, pp. 291–318.

Higgins, J. V., Bryer, M. T., Khoury, M. L. & Fitzhugh, T. W. (2005). A freshwater classification approach for biodiversity conservation planning. *Conservation Biology*, **19**, 432–45.

Holmlund, C. M. & Hammer, M. (1999). Ecosystem services generated by fish populations. *Ecological Economics*, **29**, 253–268.

Hunter, M. L. (1991). Coping with ignorance: the coarse filter strategy for maintaining biodiversity. In *Balancing on the Brink of Extinction*, ed. L. A. Kohm. Washington, DC: Island Press, pp. 266–81.

The Iraq Foundation (2003). *Building a Scientific Basis for Restoration of the Mesopotamian Marshlands: Findings of the International Technical Advisory Panel Restoration Planning Workshop.* The Iraq Foundation (www.edenagain.org/publications/pdfs/bldgscientificbasis.pdf).

IUCN (The World Conservation Union) (2003). *IUCN Red List of Threatened Species.* The IUCN Species Survival Commission. (www.redlist.org).

Kamdem-Toham, A., D'Amico, J., Olson, D. *et al.* (2003). *Biological Priorities for Conservation in the Guinean-Congolian Forest and Freshwater Region.* Libreville, Gabon: WWF-CARPO.

Kepner, W. G., Semmens, D. J., Bassett, S. D., Mouat, D. A. & Goodrich, D. C. (2004). Scenario analysis for the San Pedro River, analyzing hydrological consequences of a future environment. *Environmental Monitoring and Assessment*, **94**, 115–27.

Koeshendrajana, S. & Hoggarth, D. D. (1998). *Harvest Reserves in Indonesian River Fisheries.* Proceedings of the Fifth Asian Fisheries Forum – International Conference on Fisheries and Food Security Beyond the Year 2000. Chiang May, Thailand.

Kottelat, M. (1989). Zoogeography of the fishes from Indochinese inland waters with an annotated check-list. *Bulletin Zoologisch Museum Universiteit van Amsterdam*, **12**, 1–54.

Kottelat, M. & Whitten, T. (1996). *Freshwater Biodiversity in Asia with Special Reference to Fish.* World Bank Technical Paper No. 343. Washington, DC: The World Bank.

Master, L. L., Flack, S. R. & Stein, B. A. (1998). *Rivers of Life: Critical Watersheds for Protecting Freshwater Biodiversity*. Arlington, VA: The Nature Conservancy.

Moller, H. S. (1999). *Restoration as an Element of National Planning for Wetland Conservation and Wise Use. The Ramsar Convention.* (http://ramsar.org/cop7_doc_17.4_e.htm).

Moyle, P. B. & Yoshiyama, R. M. (1994). Protection of aquatic biodiversity in California – a 5-tiered approach. *Fisheries*, **19**, 6–18.

Navarro, G. & Maldonado, M. (2002). *Geografia Ecologica De Bolivia: Vegetacion Y Ambientes Acuaticos*. Bolivia: Centro de Ecologia Simon I. Patino, Departamento de Difusion.

Nel, J., Maree, G., Roux, D. *et al.* (2004). *National Spatial Biodiversity Assessment 2004: Technical report. Volume 2: River Component.* CSIR Report Number ENV-S-I-2004-063. Stellenbosch, South Africa: Council for Scientific and Industrial Research.

Nilsson, C., Reidy, C. A., Dynesius, M. & Revenga, C. (2005). Fragmentation and flow regulation of the world's large river systems. *Science*, **308**, 405–8.

Norris, R. H. & Thoms, M. C. (1999). What is river health? *Freshwater Biology*, **41**, 197–209.

Noss, R. F., Carroll, C., Vance-Borland, K. & Wuerthner, G. (2002). A multicriteria assessment of the irreplaceability and vulnerability of sites in the Greater Yellowstone Ecosystem. *Conservation Biology*, **16**, 895–908.

Oberdorff, T., Guegan, J. F. & Hugueny, B. (1995). Global scale patterns of fish species richness in rivers. *Ecography*, **18**, 345–52.

Olson, D. M. & Dinerstein, E. (2002). The Global 200: priority ecoregions for global conservation. *Annals of the Missouri Botanical Garden*, **89**, 199–224.

Olson, D. M., Dinerstein, E., Canevari, P. *et al.* (1998). *Freshwater Biodiversity of Latin America and the Caribbean: A Conservation Assessment.* Washington, DC: Biodiversity Support Program.

Possingham, H., Ball, I. & Andelman, S. (1999). Mathematical models for identifying representative reserve networks. In *Quantitative Methods for Conservation Biology*, eds. S. Ferson & M. A. Burgman. New York: Springer-Verlag, pp. 291–305.

Pressey, R. L., Hager, T. C., Ryan, K. M. *et al.* (2000). Using abiotic data for conservation assessments over extensive regions: quantitative methods applied across New South Wales, Australia. *Biological Conservation*, **96**, 55–82.

Pringle, C. M. (2000). River conservation in tropical versus temperate latitudes. In *Global Perspectives on River Conservation: Science, Policy and Practice*, eds. P. J. Boon, B. R. Davies & G. E. Petts. Chichester, UK: John Wiley, pp. 371–384.

PROBIO (Projeto de Conservação e Utilização Sustantavel da Diversidade Biologico Brasileira) (1999). *Ações Prioritarias Para a Conservação da Biodiversidade do Cerrado e Pantanal.* p. 27.

Ramsar Convention on Wetlands (2002). *Principles and Guidelines for Wetland Restoration. The Ramsar Convention* (www.ramsar.org/key_res_viii_16_e.htm).

Reis, R., Kullander, S. & Ferraris, C. (eds.) (2003). *Check List of the Freshwater Fishes of South and Central America.* Porto Alegre, Brazil: EDIPUCRS.

Revenga, C. & Kura, Y. (2003). *Status and Trends of Biodiversity of Inland Water Ecosystems.* Technical Series No. 11. Montreal, Canada: Secretary of the Convention on Biological Diversity.

Revenga, C., Murray, S., Abramovitz, J. & Hammond, A. (1998). *Watersheds of the World: Ecological Value and Vulnerability.* Washington, DC: World Resources Institute.

Revenga, C., Brunner, J., Henninger, N., Kassem, K. & Payne, R. (2000). *Pilot Analysis of Global Ecosystems: Freshwater Systems.* Washington, DC: World Resources Institute.

Revenga, C., Campbell, I., Abell, R., de Villiers, P. & Bryer, M. (2005). Prospects for monitoring freshwater ecosystems towards the 2010 targets. *Philosophical Transactions of the Royal Society B – Biological Sciences*, **360**, 397–413.

Roux, D., de Moor, F., Cambray, J. & Barber-James, H. (2002). Use of landscape-level river signatures in conservation planning: a South African case study. *Conservation Ecology*, **6**, 6 (www.consecol.org/vol6/iss2/art6/).

Shumway, C. A. (1999). *Forgotten Waters: Freshwater and Marine Ecosystems in Africa. Strategies for Biodiversity Conservation and Sustainable Development.* Alexandria, VA: Global Printing, Inc.

Silva, X. & Guevara, M. (2003). *Catalyzing Conservation Action in Latin America: Identifying Priority Sites and Best Management Alternatives in Five Global Significant Ecoregions.* Final report Prepared for United Nations Environment Programme with support from the Global Environment Facility (GEF/1010-00-14). The Nature Conservancy & NatureServe.

Smith, R. K., Freeman, P. L., Higgins, J. V. *et al.* (2002). *Priority Areas for Freshwater Conservation Action: A Biodiversity Assessment of the Southeastern United States.* Arlington, VA: The Nature Conservancy.

Sowa, S. (2004). *The Aquatic Component of Gap Analysis: A Missouri Prototype Final Report.* Missouri Resource Assessment Partnership, University of Missouri. Submitted to The United States Department of Defense Legacy Program, Project Numbers: 981713 and 991813. 30 September 2004.

Terneus, E., Cárdenas, A., Calles J. A. & Beltrán, K. (2004). *Informe Final del Proceso de Evaluación Ecorregional del Proyecto Pacífico – Ecuatorial (Componente de Agua Dulce).* Quito, Ecuador: The Nature Conservancy.

Tharme, R. E. (2003). A global perspective on environmental flow assessment: emerging trends in the development and application of environmental flow methodologies for rivers. *River Research and Applications*, **19**, 397–441.

Thieme, M. L., Abell, R., Stiassny, M. L. J. *et al.* (2005). *Freshwater Ecoregions of Africa and Madagascar: A Conservation Assessment*. Washington, DC: Island Press.

United Nations (2001). *Road Map Towards the Implementation of the United Nations Millennium Declaration*. Report A/56/326. Office of the Secretary-General, United Nations General Assembly.

USAID (United States Agency for International Development) (2003). *Biodiversity Conservation: A Report on USAID's Biodiversity Programs in Fiscal Year 2002*. Washington, DC: USAID.

Valbo-Jørgensen, J. & Poulsen, A. F. (2000). Using local knowledge as a research tool in the study of river fish biology: experiences from the Mekong. *Environment, Development and Sustainability*, **2**, 253–76.

van der Linde, H., Oglethorpe, J., Sandwith, T., Snelson, D. & Tessema, Y. (2001). *Beyond Boundaries: Transboundary Natural Resource Management in Sub-Saharan Africa*. Washington, DC: Biodiversity Support Program.

Vannote, R. L., Minshall, G. W., Cummins, K. W., Sedell, J. R. & Cushing, C. E. (1980). The River Continuum Concept. *Canadian Journal of Fisheries and Aquatic Sciences*, **37**, 130–37.

Wasson, J.-G., Barrera, S., Barrere, B. *et al.* (2002). Hydro-ecoregions of the Bolivian Amazon: a geographical framework for the functioning of river ecosystems. In *The Ecohydrology of South American Rivers and Wetlands*, ed. M. E. McClain. Wallingford, UK: International Association of Hydrological Sciences, pp. 69–91.

Weitzell, R. E., Khoury, M. L., Gagnon, P. *et al.* (2003). *Conservation Priorities for Freshwater Biodiversity in the Upper Mississippi Basin*. NatureServe and The Nature Conservancy.

Wishart, M. J. & Davies, B. R. (2003). Beyond catchment considerations in the conservation of lotic biodiversity. *Aquatic Conservation: Marine and Freshwater Ecosystems*, **13**, 429–37.

Wishart, M. J., Davies, B. R., Boon, P. J. & Pringle, C. M. (2000). Global disparities in river conservation: 'First World' values and 'Third World' realities. In *Global Perspectives on River Conservation: Science, Policy and Practice*, eds. P. J. Boon, B. R. Davies & G. E. Petts. Chichester, UK: John Wiley, pp. 353–69.

World Bank (2001). *Making Sustainable Commitments – An Environment Strategy for the World Bank*. Report No. 23084. Washington, DC: The World Bank.

WWF (2004). *Dam Right: Rivers at Risk*. Gland, Switzerland: WWF International.

WWF (2005). *A Biodiversity Vision for the Amazon River and Floodplain Ecoregion*. Brasilia, Brazil: WWF.

Yoffe, S., Fiske, G., Giordano, M., Giordano, M. *et al.* (2004). Geography of international water conflict and cooperation: data sets and applications. *Water Resources Research*, **40**, 1–12.

13 · *Conclusions*

CATHERINE M. PRINGLE AND
PHILIP J. BOON

The protection of freshwater ecosystems is a formidable challenge in countries throughout the world. As pointed out by Pringle and Withrington (Chapter 3), and illustrated in examples within this book, trade-offs between human use of water resources and protection of the natural functioning of fresh waters in different countries are determined by many factors, reflecting differences in the history of human settlement, socio-economics, politics, geographic extent, climate and many others. This book primarily focuses on a 'developed country perspective' contrasting views and approaches of the United States and the United Kingdom (Chapters 1–8), but also including additional European insights from Sweden (Chapter 9), along with comparative chapters on Australia (Chapter 10) and South Africa (Chapter 11). As these chapters indicate, while freshwater conservation and evaluation may be implemented quite differently, the guiding principles of conservation biology are often similar.

In both the USA and the UK, conservation is achieved through the efforts of national and international agencies, non-governmental conservation organizations (NGOs), and academics (Boon & Pringle, Chapter 2; Pringle & Withrington, Chapter 3). In the UK, the conservation movement has developed through mainstream government agencies over more than half a century, influenced strongly by international European legislation. The USA, while strongly influenced by national legislation, is clearly less influenced by international legislation. Moreover, in the USA, NGOs have played a key role in the development of freshwater conservation for natural values, ranging from developing conceptual paradigms to both collaborating with, and litigating against, government agencies. While NGOs are important in conservation in the UK, they do not play the central role that they do in the USA (Pringle & Withrington, Chapter 3).

As pointed out by Higgins and Duigan (Chapter 4), the USA and the UK have collectively taken comprehensive and proactive approaches to setting priorities for biodiversity conservation. These approaches include

Assessing the Conservation Value of Fresh Waters, ed. Philip J. Boon and Catherine M. Pringle. Published by Cambridge University Press.

prioritizing the last of the least (to protect against future loss of species and habitats) and *incorporating the best of the rest* to keep more species and habitats from becoming at risk (Higgins & Duigan, Chapter 4). While numerous examples of well-coordinated conservation efforts for specific regional or site-planning activities can be found in both the countries, a greater level of coordination to generate comprehensive national plans would enhance the effectiveness of conservation planning (Chapter 4).

In many cases, of course, long-term strategic planning for conservation is not the issue. Instead, when proposed activities within catchments threaten vulnerable freshwater habitats and species, the focus may shift to seeking immediate protection through development control systems (Frissell & Bean, Chapter 5). There are broad similarities in the way that these are used in the USA and the UK, but there are differences as well. Both countries have adopted the general principles of environmental impact assessment (EIA), supported by legislation such as the National Environmental Policy Act in the USA, and the European EIA Directive and SEA Directive in the UK. Yet, by contrast, the emphasis on NGO and citizen action to confront developments affecting freshwater ecosystems is far greater in the USA.

Where damage has occurred, and rivers and lakes have become degraded, restoration may become an option (Langford & Frissell, Chapter 6). Both in the USA and the UK, the past 20 years have seen a significant increase in the number and size of freshwater restoration schemes, especially for rivers. The incentives for restoration are many and varied, including human safety or welfare, economic production or public amenity, and ecological or conservation aims are not often the primary motivation for restoration. Langford and Frissell (Chapter 6) conclude that 'despite all these sources of information and the plethora of literature proposing methods and approaches, it is likely that most large-scale freshwater 'restoration' projects in the near future will be proposed and implemented on the basis of political expediency or public safety rather than on grounds of conservation or ecology.' Nevertheless, when ecological restoration is planned and implemented, it is vital that suitable evaluation techniques are used both at the planning stage and after completion to monitor the success of the scheme.

Methods used in the UK and the USA for assessing river (Boon & Freeman, Chapter 7) and lake (Duker & Palmer, Chapter 8) conservation share a common emphasis on classifying habitats and collecting biodiversity information in order to protect biological distinctiveness. In both countries, goals of river management and river conservation have moved closer together. The ultimate goal of integrated catchment management seems to be climbing in the political agenda of both countries. Biological monitoring, beyond traditional water chemistry monitoring of a few selected chemical parameters, has clearly become more important in both countries over the last two to three decades. Biological monitoring in rivers has moved, in some cases, beyond an emphasis on macroinvertebrates to other organisms (e.g. plants and fish). Similarly, the use of biodiversity indicators in lakes is beginning to move beyond an emphasis on plants and birds. There is an increasing awareness of the importance of understanding and protecting geomorphological processes that shape river habitats, as well as the ecological processes that structure freshwater communities (Chapter 7). It is encouraging to note that volunteer monitoring programmes and NGOs are playing increasingly important roles in river and lake biodiversity and health assessments in both the USA and the UK, providing models that may be transferable to other countries (Chaper 8). However, the task of monitoring the 'ecological health' of a freshwater ecosystem is not always synonymous with evaluating its conservation importance, and assessment systems need to be aware of the similarities and differences.

Chapter 9 (Willén) provides a further illustration of the approaches taken in Europe for evaluating fresh waters for conservation, showing how standard methods for biological monitoring and traditional ways of assessing conservation value are beginning to move closer to each other. In Sweden, the development of a conservation evaluation tool (System Aqua) has demonstrated the importance of using techniques that are not only repeatable and robust, but which provide results that can be readily understood by planners and policy makers as much as by scientists. The need for these sectors of society to work together is well illustrated by the increasing influence of European directives in freshwater management. It is worth noting that one of the perceived advantages of System Aqua is its capacity for continual adaptation to meet the demands of the EC Water Framework Directive (WFD), as well as providing an objective way of assessing nature conservation value.

The transatlantic comparisons given in this book are complemented by the final three chapters which take the themes in Chapters 2–8 and provide short summaries of each one for three contrasting regions of the world. Chapter 10 (Nevill & Boulton) covers Australia, where the first-world paradigms of freshwater conservation share many similarities with those in the USA and the UK. All state governments, and the national government, are committed to the principle of conserving aquatic biodiversity through protecting representative ecosystem types and by careful land management outside protected areas. Yet, very limited progress has been made in achieving these aims. For this type of approach to succeed, methods for selecting areas for protection must be in place. Nevill and Boulton (Chapter 10) propose that using techniques of 'systematic conservation planning' is likely to be more fruitful in determining conservation priorities than using scoring approaches.

Chapter 11 (O'Keeffe & Thirion), on South Africa, bridges the gap between the developed-world attitudes to nature conservation in the preceding chapters, and the contrast that Chapter 12 (Abell & Bryer) provides. The socio-political history of South Africa, coupled with the rapid changes in society over the last 15 years, has led to an interesting mixture of approaches for assessing the conservation value of rivers and lakes. Some, such as the River Conservation System (RCS) devised in the late 1980s (O'Keeffe *et al.*, 1987) or the method for assessing habitat integrity developed by Kleynhans (1999), share much in common with systems used in the UK and the USA. Indeed, the RCS provided the impetus for the production of SERCON in the UK (see Chapter 7). However, O'Keeffe and Thirion (Chapter 11) stress that in South Africa conservation based on values such as rarity or species diversity will only be successful if these attributes are seen also as having economic value.

It is perhaps the stress on socio-economics that (at least until recently) differentiates attitudes in developing countries from those in the developed world. While this book largely focuses on the latter, Chapter 12 stresses the marked disjunct between conservation and evaluation of fresh waters in developing versus developed countries. As pointed out by Abell and Bryer (Chapter 12),

> Developing country assessments are now creating a new dialect within the larger language of freshwater conservation assessments ... Biodiversity values are often high, but data describing them are few, and pressures of impending development may necessitate making conservation recommendations without the benefit of even basic biological information. With highly constrained financial resources, the motivation for developing nations to invest in freshwater conservation understandably tilts more toward the provision of ecosystem goods and services than toward biodiversity protection.

Yet, this is no longer only a developing-world view. The subject of 'ecosystem goods and services' is now entering the vocabulary of some government departments and conservation bodies in the UK, USA and other developed countries – where the importance of conserving nature 'for its own sake' is now receiving less prominence than the economic gains that nature can bring to human society.

It was not our intention in this book to provide a global coverage of the subject, and those parts of the world that are rapidly developing are not discussed. Nevertheless, we must acknowledge the increasing importance of globalization and the economic influence of China and other Asian countries. This will surely play an increasing role in the further development of world views on freshwater conservation, which have previously been dominated by European and western views.

In summary, effective freshwater conservation requires greater integration of survey and inventory approaches with evaluation and conservation planning to ensure practical conservation management solutions. The challenge of freshwater conservation in the new millennium will require significant further efforts to bring policy and practice into line with our scientific understanding of how rivers and lakes work. This is clearly a daunting task given the importance of the landscape-scale and global processes that maintain freshwater ecosystems and the extent of the changes brought about by expanding human populations.

Index